Quest for Excellence Through Globalisation

Quest for Excellence Through Globalisation

P.K. Perumal, D.Litt.

NEW DAWN PRESS, INC.
USA• UK• INDIA

NEW DAWN PRESS GROUP

Published by New Dawn Press Group
New Dawn Press, Inc., 244 South Randall Rd # 90, Elgin, IL 60123
e-mail: sales@newdawnpress.com

New Dawn Press, 2 Tintern Close, Slough, Berkshire, SL1-2TB, UK
e-mail: sterlingdis@yahoo.co.uk

New Dawn Press (An Imprint of Sterling Publishers (P) Ltd.)
A-59, Okhla Industrial Area, Phase-II, New Delhi-110020
e-mail: info@sterlingpublishers.com
www.sterlingpublishers.com

Quest for Excellence through Globalisation
© 2006, Dr. P.K. Perumal
ISBN 1-84557-592-X

All rights are reserved. No part of this publication may be reproduced, stored in a retrieval system or transmitted, in any form or by any means, mechanical, photocopying, recording or otherwise, without prior written permission of the publisher.

PRINTED IN INDIA

Dedicated to

CHEVALIER DR. K. THIAGARAJAN, D.sc.,
Secretary General
United Writers Association, Chennai

A Potent Source of Perennial Inspiration

Acknowledgements

I am highly indebted to a group of *management wizards* whose previous works/research articles etc., have been greatly instrumental in shaping my line of thinking. Especially the works of stalwarts like Mr. Tom Peters, Prof. Peter Drucker, Prof. Michael Porter, Prof. Kenichi Ohmae, Prof. Warren Bennis, Prof. Rosabeth Moss Kanter, Dr. Gary Hammel and Mrs. Joan Magretta and Nanston containing useful guidelines on *futuristic thinking* required on issues of 'Globalisation' have helped me a great deal. On the basis of their sparkling ideas, I have constructed a suitable *strategic approach* for achieving *excellence* as well as *world class status* by a *modern firm*.

When the book was in the germinal stage, my conversations with Chevalier Dr. K. Thiagarajan, Secretary General of the United Writers Association, Chennai, helped me much to deepen my thinking and gain several remarkable insights.

I am also indebted to Major S. Lakshmanan, Executive Director of Madras Management Association, Chennai for writing a foreword to the book.

I am enormously grateful to the wonderful cooperation extended by the management/staff of British Council Library, and the USIS Library, Chennai.

My thanks go to Mr. Murali and Mr. Venugopal of M/s. Sowmya Enterprises, for their help in editing the publication.

Finally, my sincere appreciation and grateful thanks, to Mr. S.K.Ghai, Managing Director Sterling Publishers Pvt. Ltd., who encouraged me to write this book.

Dr. P.K. Perumal

Foreword

Though there is a rich fund of sources of reference on the subject of 'Globalisation', only a few publications offer pragmatic, rejuvenating ideas particularly useful for young junior executives and research students.

Here is a publication presented by Dr. P.K. *Perumal* management expert and economist, exclusively meant for the benefit of junior executives associated with firms, young entrepreneurs and research fellows. It explores several dimensions of the managerial framework of a firm and offers wonderfully fine insights into improving the 'organisational performance' of a 'modern firm' aspiring for 'world class status'.

Taking into account the highly competitive contemporary global business scenario, this book lays much accent on the need for acquiring a 'paradigm shift' in approach connected with: strategy formulation, intellectual capital formation, specifically tailored training programmes and consistently offering customised products/services of high quality by any 'modern firm' keenly desirous of achieving 'world class status'.

Since 'quality of life' has received greater emphasis in the 21st century socio - economic milieu, firms and organisations could successfully thrive only on the basis of 'genuine value additions' they provide towards the welfare of the society as a whole.

I wholeheartedly recommend this publication for any one who is keenly interested in getting a refreshing exposure to the management tools and techniques so essential for achieving 'Excellence' by a modern firm in the 21st century milieu.

Major S. Lakshmanan
Executive Director

Madras Management Association (MMA),
Chennai

Preface

This book has been written with the objective of enabling the young junior executives, managers as well as budding young enterpreneurs to gain an insight into the cardinal management principles. It offers useful guidelines to young managers on how strategic decisions could be meticulously crafted by an 'aspiring firm' functioning in a turbulent global economic scenario.

While functioning in a highly competitive environment, a 'modern firm' keenly aspiring to achieve world class status should indeed strive for total organisational excellence by understanding each activity and involving each individual at each level.

It is an acknowledged fact that effective leadership plays a key role in organisational excellence. Accordingly, 'management as a whole' should make concerted endeavours towards reaching world class performance through obsessional commitment by ensuring the turnout of consistently high quality products, while conforming to 'international standards' by imparting special training to a core group of highly talented people in consonance with organisational needs.

Besides, the aspiring modern firm should avoid internally myopic operations which are deleterious for the healthy growth of the firm in the long run. 'Team spirit' and 'creativity' assume great significance in the modern organisational framework.

Organisational excellence can not be achieved without a radical departure from orthodox techniques and procedures, and a change in the 'mindsets' of all those, associated with the management pool would be absolutely essential to achieve resounding success.

Besides discouraging downsizing, the book lays considerable emphasis on 'core competencies' and providing added value to its devoted and loyal customers. While dynamic changes are taking place in a highly competitive global market place, the need behind undertaking constant innovations, customised production and endeavours towards exceeding customer expectations has been stressed.

In order to gain the required competitive edge over other firms, the significance of job enrichment, speed, accuracy and efficiency in delivering products and services for the modern firm can hardly be ignored.

In addition, several modern concepts such as 'six sigma production', 'elimination of total wastage', 'competency mapping' and 'creating a pool' for grooming highly talented people to reach higher positions have been given sharp focus.

Chapters I and VIII particularly provide an overview of and insights into aspects of achieving 'Excellence by a global firm'.

It is my fervent hope that this publication would stimulate considerable interest among junior executives/managers, keenly interested in 'breakthrough strategies' required by a modern firm aspiring to reach global status.

Chennai

Dr. P.K. Perumal

Contents

	Foreword	*vii*
	Acknowledgements	*ix*
	Preface	*xi*
	List of Figures and Tables	*xiv*
1.	Understanding Globalisation	1
2.	Strategic Thinking: Key to Global Competitiveness	18
3.	Technological Competence	42
4.	Customer Focus and Global Marketing Strategy	67
5.	Organisational Changes	90
6.	Quality as a Cohesive Policy for Achieving Excellence	114
7.	Harnessing Managerial Dynamism: Vision for 21st Century	144
8.	Strategy for Excellence in a Borderless World	171
	Bibliography	226
	Index	230

List of Figures

1.1	Flexibility and Competitive Advantage	17
2.1	Formal Steps Involved	23
2.2	Modern Firms - Strategic Management	25
2.3	Opportunity Staircase for the Modern Firm	27
2.4	Product Strategy - Analysis	36
2.5	Pair of Scissors Analogy	41
3.1	An Ideal Manufacturing Continuum	55
4.1	Greater Customer Satisfaction	76
4.2	Kano's Model	76
4.3	Model on Core Business Elements	78
4.4	Focus on Outer Rings of the Total Product	79
4.5	Decisions to be taken by the Global Firm	83
5.1	Vision Mission and Rover Group's Success Factors	95
5.2	Emergence of World Class Quality	99
5.3	Actual Performance Improvement Obtained in Relation to Intended (Breakthrough) Improvement	104
5.4	When Kaizen Approach is Adopted	106
5.5	Essential Features of Breakthrough and Continuous Improvements	107
5.6	Competitive Advantage	108
5.7	Strategic Staircase	111
5.8	Process Management Tool Box	111
6.1	Principles Determining Quality/Performance - Standards	116
6.2	Five Salient Absolutes of Crosby	119
6.3	Crosby's 14 Step Programme	122
6.4	PDCA Cycle	124
6.5	The Seven Deadly Sins	125
6.6	A Sample Control Chart	126
6.7	Deming's Principles for Transformation	127
6.8	Ten Benchmarks Essential to Achieve Total Quality Success	129

6.9	Quality Trilogy of Joseph Juran	130
6.10	Juran's Ten Steps to Continuous Quality Improvements	131
6.11	Process Improvement Initiatives	136
6.12	Six - Sigma Quality: The Bell Curve	139
7.1	Double Loop Learning	157
7.2	Policy Deployment Process	160
7.3	Never - ending Improvement Helix	162
7.4	SWOT Summary	165
7.5	Conversion into Intangible Capital	169
8.1	Michael Porter's Competitive Forces – Model	175
8.2	Ideals and Ideologies Necessary for a Global Firm	179
8.3	Comparison of Traditional and Global Mindsets	180
8.4	Personal Characteristics/Implications for Global Mindset	181
8.5	Leap Frogging Globalisation Pathways	182
8.6	Characteristics of a Multinational Company and a Global Firm	187
8.7	An Useful Comparison of Multinational and Global Organisations	188
8.8	An Ideal Model of HRD Process	193
8.9	Dr. Maslow's Hierarchy of Needs	195
8.10	The Star Model: To Achieve Significant Change	198
8.11	Prof. Michael Porter's "Value Chain Model"	199
8.12	The Causal Model of David Maister	207
8.13	'Management' Vs 'Leadership'	209
8.14	Split Brain Theory – Essential Features	211
8.15	Cultural Contrasts/Values Appealing	222

List of Tables

1.1	Malcolm Baldridge National Quality Award	15
3.1	Advances Effected in Recent Decades	44
5.1	Firms overtaken by Rivals	101
5.2	Firms which had Reinvented	101
5.3	Intangible Qualities	113
7.1	Blending of Certain Characteristics	159
7.2	SWOT Analysis	164

1

Understanding Globalisation

"Far away there in the Sunshine are my highest aspirations; I may not reach them, but I can look up and see their beauty, believe in them, and try to follow where they head."
—**Louisa May Alcott**
American Writer/Author

I. INTRODUCTORY REMARKS

Globalisation has assumed a distinctive theme in the present decade. Several 'management gurus' have identified it as the foremost imperative in the formulation of national strategies for rapid development. It is also widely believed that only globalisation holds the 'key' to future world economic development. As such, it is also considered inevitable and irreversible. However, there are certain countries and segments of population which regard it with hostility, since they are inclined to think that globalisation may lead to increasing inequality within a country, and between nations threatening employment opportunities and living standards and would thwart social progress.

While designing suitable policies for rapid growth, it becomes utmost essential for a country to tap the gains of 'Globalisation Process' by eliminating to the maximum possible extent, different types of risks involved in the transitional phase.

Quite a good number of countries have already become integrated with the global economy. These countries are experiencing faster economic growth in addition to reducing poverty levels. Especially East Asian countries have achieved greater prosperity and dynamism by adopting outward oriented policies. Thus areas classified as poorest in the world 40 years ago have been transformed; much progress has been perceptible in adopting democratic principles as living standards started rising.

On the other hand, several countries especially in Latin America and Africa have demonstrated their capability for achieving success in the economic front by following inward oriented policies, these are special cases.

Macro Economic Implications

Of course, globalisation brings about in its wake several risks:
- Risk factors associated with highly volatile capital-movements
- Risks arising out of environmental degradation, and certain segments remaining poor as benefits of globalisation do not percolate to their level.

The foremost objective of a healthy developing country should be 'extermination of poverty' on a large-scale basis. Suitable policy changes have to be carried out in order to build healthy economies with stronger financial systems. It is also imperative that the developing countries catch up with the advanced countries of the west enjoying higher standards of living.

Some important questions which baffle macro economic policy makers and economic administrators in several countries centre on the issues:

(i) Whether globalisation would enhance inequalities, or whether it could considerably help reduce poverty on a large scale basis?

(ii) Whether integration with the global economy would make a country's economy vulnerable to instability?

These are some of the challenging issues, which have to be sensibly tackled. However, humanitarian sensitivity to tackle problems of people enjoying abnormally low standards of living has considerably affected the 'mind sets' of government policy makers of several countries.

Meaning of Globalisation

Economic globalisation could be considered as a historical process where the results of human innovations and technological progress achieved by countries in various fronts produce considerable impact on all segments of economy and society and transform the lives of the people all over the globe.

On account of this, economies around the globe achieve 'greater integration' as outlined below:

(i) Trade and financial flows are significantly augmented.
(ii) Greater movement of people especially labour force and of technical knowledge across international borders takes place.
(iii) Major changes could be perceived in political/economic/cultural fields.

In specific countries markets as such are instrumental in promoting efficiency through competition; specialisation allows people and the economy to focus on what they do is the best.

As regards global markets, they throw open glamourous opportunities which could be tapped effectively, only by those countries/organisations-who could respond instantaneously to global requirements.

This would involve tailoring a suitable policy framework that would accentuate greater exports, latest technological know-how, cheaper imports and so on.

Accordingly, organisational renewal in different industry groups of a developing economy must be able to cope with the changing political scenario when several new re-alignments are taking place.

At Micro Level

CEOs/Managers of several leading firms all over the globe are increasingly making decisions, which would enhance cross border flows on account of the flexible policies being adopted by several governments.

Indeed, globalisation by and large would trigger 'hydra headed growth' which would involve altogether a 'new paradigm' of enticing opportunities to adventurous entrepreneurs.

However, the acquisition of competitive edge by any healthy firm in the 21st century milieu would be highly dependent on:

(i) Nurturing growth intoxicated-mindset, especially among CEOs/Managers.

(ii) Capability to mobilise talents suited to global operations.

(iii) On capitalising the vast intellectual capital available as John Welch, Junior CEO of General Electric Co. has observed: "which covers willingness to share knowledge, internal knowledge creation, and effective application for greater benefits."

But the very organisational structure and the strategic behaviour of the modern firm, should well be attuned to the international macro economic management needs.

Firms, which are averse to risk taking cannot forge ahead in the 21st century milieu. When competition becomes intensified throughout the globe, an aspiring firm should make every possible endeavour to adhere to core-competencies which would fulfil world class standards. Some of the tenets which the modern firm should meticulously follow in order to survive are:

(i) Manufacturing products conforming to quality standards expected by consumers-covering product quality, services offered and management while producing consistently best quality products.

(ii) Fulfilling the need for constant upgradation by following bench marking ethics, and periodically comparing the performance of an unit's products/processes with world leaders producing similar products.

(iii) Carrying out improvements in several fields such as:
   (a) Speed/Accuracy in executing orders
   (b) Adopting innovative techniques
   (c) Improving training facilities and providing adequate employee health care and safety etc.
   (d) Correct invoicing/pricing etc.

Every firm should formulate a strategic plan for achieving success in the global front, keeping in view the need for earning substantial profits.

Areas which require greater improvement must be well identified and prioritisation should be adopted in implementation.

(iv) When *core competencies* are found inadequate, the firm could very well consider:
   (a) Outsourcing
   (b) Making strategic-alliances
   (c) Mergers/acquisitions so as to tackle the emerging challenges.

Problems Confronted
General Characteristics

During globalisation invariably world trade and financial markets of several countries become 'integrated.'

However in this process of integration, quite a good number of countries have shown only little progress:

(i) Especially per capita income in countries of Asia and Africa has actually declined; and these countries are struggling hard to catch up with the changing scenario.

(ii) Considering the actual composition of exports, substantial changes have taken place in recent years, and only those countries exporting manufacturing goods are thriving well. Poorest countries, especially those exporting primary commodities such as food based items and raw materials and so on, have been badly hit on account of decline in their share of exports.

(iii) In addition, there are considerable adverse changes in regard to the composition of private capital flows especially foreign direct investment to certain developing countries in Asia.

(iv) Migration of skilled workers mostly occurs between developing countries. However, in regard to developing countries, those who go abroad get specialised skills, and demand more wages on a par with what they used to get in the advanced nations after returning to their home countries.

Several countries in transition from planned to market economies are making every possible endeavour to get integrated with the global economy, and in this regard, except Poland and Hungary, several countries including some of the former Soviet Union are confronted with long term structural and institutional issues and are making slow progress in this direction.

(v) Problem of maintaining macro economic stability in the short run.

In a world of integrated markets, most of the developing countries are affected by *'volatile capital flows'*. The risk factors associated with policies meant to promote financial stability are quite high. Countries may be confronted with problems in implementing *wage increases and price make ups* which may make, a specific country-*uncompetitive*.

If a country takes too sanguine a view of the continual inflow of capital movements from other countries and adopt flexible policies, it may be exposed to risks on account of drastic changes announced by other countries for withdrawal of capital suddenly in response to unforeseen international pressures.

Certain Important Implications

(i) Management wizard: Peter Drucker has stated that spectacular bursts in economic activity which took place in the middle of the 19th century could largely be attributed to the *'introduction of rail roads'* in several countries, which brought about the Industrial Revolution. The mental geography of people had undergone a rapid change on account of this innovation and human beings had acquired greater mobility with expanding intellectual horizons.

In a similar way recently in the global scenario, *E-Commerce* has brought about the information revolution akin to the Industrial Revolution that took place in the 19th century.

While the rail road created a new mental geography to master the distance in the 19th century, in the recent decade, E-Commerce has almost eliminated distance, and has established a 'borderless world'. As a result, it has become imperative for every business to become globally competitive in order to survive.

All the manufacturers, whether they cater for local markets, or regional markets or international markets have no alternative except to tackle global challenges in the most effective manner possible.

(ii) New technologies and new industries are rapidly emerging. Rapid strides in information technology, induction of computer data processing, and the Internet will certainly accentuate the emergence of new industries, across the globe during the next decade.

Bio-technology and fish farming have assumed unusual fascination in recent years. Besides, insurance against the risk of foreign exchange exposure has assumed crucial significance. New knowledge based industries are gradually multiplying. However, effectively running these institutions depend mostly on the capability of the modern firm to increasingly attract, retain and motivate knowledge workers by offering greater incentives and social recognition.

Appropriate Policy Measures

(i) While achieving increased integration in the financial sphere on account of globalisation, several problems may be confronted in managing the economic affairs-such as limitations imposed in a government's choices of tax rates, exercising freedom in pursuing monetary policies, exchange rate policies and so on. However, by and large, from the point of view of achieving sustainable economic growth rates, greater social progress, and low inflation rates, globalisation would certainly prove highly beneficial to most of the countries in the long run.

(ii) Governments can thrive only by adopting sound economic policies. Private enterprises should receive a boost and should be encouraged to take up suitable product lines after carefully analysing the risks involved. Besides, every country should coordinate in the endeavours to create a sound international financial architecture in order to stabilise the international capital flows and to make impressive records of economic performance.

Factors to be given accent

Countries integrating with global markets should follow certain principles in order to succeed:
- (*i*) Adopting suitable policy framework to achieve macro-economic stability.
- (*ii*) Financial soundness
- (*iii*) Ensuring transparency in transactions
- (*iv*) Good governance

Shortcomings to be avoided
- (*i*) Making investment decisions too quickly, without proper appraisal of risk.
- (*ii*) Inadequate feed back mechanisms in regard to latest developments in major financial centres.
- (*iii*) Tendency to adopt herd behaviour in specific markets.
- (*iv*) Sudden shift in investment sentiments and rapid movements of short-term finance inside and outside the countries.

As such, absence of financial safety nets at the national level are deleterious to make rapid progress. At the international level – the IMF, WTO and other financial organisations should take up suitable control mechanisms to avoid the impact of any crisis.

Further Observations

Globalisation, by and large, has so far benefited most of the advanced countries, and to some extent a few developing countries.

Several economists/management experts have expressed the view that there is a growing tendency for widening the gap between the 'high income countries' and 'low income countries'.

Quite a good number of citizens still languish in abject poverty in miserable conditions, with little employment opportunities.

As such, countries having low incomes and low standards of living, are mostly struggling hard to integrate with the global economy because of their pessimistic laissez faire policy frame works, focusing attention mostly on short term advantages. Several such societies which have their cultural identity and remain isolated without absorbing the boundless flow of new ideas and innovative technologies fail to 'integrate' with the global economy, and remain in perpetual poverty. In this regard the international community, besides strengthening the international financial system, should endeavour to motivate and help:

(i) Expansion of trade

(ii) Provision of substantial aids, so as to impart dynamism and enthuse them to reconstruct 'perfect, liberated and egalitarian societies'.

The poorest countries, accustomed to lackluster performance, should exterminate poverty, and should have access to globalisation only by suitable reconstruction surveillance and stabilisation measures and these economies could very well be 'integrated' with the global economy so that the equilibrium of planet earth could also be realised.

II. CREATIVE EXCELLENCE

Creativity flourishes in an organisation, where top level managers precisely understand the kind of managerial practices that would foster creativity. The managers should opt for work place practices and conditions which would stimulate the following important characteristics with which each individual has been endowed:

(i) Adequate knowledge and expertise covering intellectual, technical and procedural aspects.

(ii) Special skills for creativity possessed by individuals, and their flexibility and imaginative approach.

(iii) Intrinsic motivation-whether the individual has inherent passion to solve problems in an work environment which is conducive.

'Creativity' could very well be enhanced when managers recognise these inherent traits, and provide the supportive environment.

Besides, the foremost task of a highly capable manager would be matching people with the most appropriate assignments. By achieving perfect matching, the manager would be able to *ignite intrinsic motivation* which would ultimately ensure that the employees' abilities are fully *stretched*.

This kind of perfect matching could be achieved only if the manager takes time and trouble to gather rich and detailed information about his employees and the available assignments. In this regard quick-fix solutions based upon urgency may not be of much help. Adequate autonomy to employees combined with setting up of *strategic goals* would be found highly beneficial.

Even while confronted with a serious problem such as *competitors activities* to launch a great product at a lower price, the manager should opt for a *sophisticated judgement* that would help him to forge ahead after carefully assessing the following factors:

(i) Marketing people with right assignments
(ii) Time and money to be allotted:
(iii) Avoiding impracticable deadlines and unusually tight time schedules.

'Creativity' necessary to reach high standards of product excellence would be killed in a process that creates distrust and people feel over-controlled. Adequate time allotted for worthwhile exploration of unique solutions, and allowance for incubation periods would greatly enhance creativity.

Building Teams to Foster 'Creativity'

While constituting a team to come up with worthwhile and creative ideas, certain factors have to be borne in mind.

(i) They should be mutually supportive in attitude with diverse backgrounds and complementary to one another.

(ii) They should comprise people of various intellectual foundations/expertise etc.

(iii) Team-mates should be able to function in a highly co-operative spirit, helping each other, and sharing the enthusiasm and excitement to achieve the *team's goal* envisaged.

Such a team of stalwarts having the right chemistry, would help to enhance expertise and creative thinking. On the other hand, it has been found by experience that homogeneous teams with similar mindset invariably achieve very little.

Creating Excellence in the Business World

We recognise 'excellence' in the works of William Shakespeare and Aldous Huxley and musical notes of Beethoven and Mozart. Similarly, we are also highly appreciative of excellence in the business world. Our minds and hearts are captivated by Rolls-Royce and Mercedez Benz cars; travel by supersonic jets, as well as in the renowned passenger ship: Elizabeth-2 would be designated as incomparable, providing the greatest personal fulfilment.

Creating 'excellence' depends mostly on:

(i) Leaders who ignite creativity

(ii) Enthusiastic innovators

(iii) Adventurous entrepreneurs and

(iv) Business executives who induce creativity.

In the 21st century milieu, *technological breakthroughs* are taking place on account of the tireless efforts of (i) *multidisciplinary teams and* (ii) *dream teams*; isolated efforts are few !

Precise market place knowledge alone could help to capture a better view of business priorities, sought by the modern firm. Today's leading business enterprise, undoubtedly requires a leading edge in regard to R and D capability.

As such technological knowledge could be 'codified' into two major categories:
- *(i)* Codified knowledge which could be protected as the intellectual property of the firm.
- *(ii)* Tacit knowledge in regard to the high technical skills and specialised competencies of industrial scientists/ employees associated with the 'modern enterprise.'

Every modern organisation should opt for a *creative work culture*, in order to turnout new products and exploit new opportunities. The scientists, engineers and executives of the modern firm should possess contagious enthusiasm to turn out new and original products, and the required services with great perfectionism.

Encouraging behavioural stereotypes and endeavours to maintain 'status quo' would only hamper creativity !

Certain important characteristic traits are essentially stressed by Dr. Parmerter and Dr. Garher in their research findings; these are:
- *(i)* The modern firm should adopt a flexible, goal oriented approach and strive to achieve always something original.
- *(ii)* Besides having 'multilevelled thinking frame' with great freshness, the modern firm should have the strong tendency to see the 'wood' and not the trees.

Considering such factors, the global firm should be able to formulate a new 'suitable gestalt-approach' which would certainly help the firm to grow on healthy lines earning substantial profits.

Management Style with Accent on Creativity

Several leading global firms have indicated that 30 to 40% of their sales are the result of products which are less than 4 years old.

The CEO/General Manager, convening important meetings should therefore *continually reinforce* the utmost need for developing new products in consonance with the latest market requirements, fashions, fads and so on. Such an approach must be ingrained in the management style of the aspiring firm.

Great enterprises such as: General Electric Co., Sony, Siemens, Hewlett Packard, Honda, Toyota, Nissan, etc., could be termed *global* since they have surpassed several firms in terms of *product excellence* and *performance standards*.

The R and D laboratories of the leading global firms permit their outstanding scientists/engineers/executives to have sufficient academic freedom to dedicate roughly 15 to 20% of their time to explore new improvements in respect of existing product lines.

Such a flexible firm would lay accent on nurturing *innovative ideas* and constantly develop new products which prove highly successful with their impressive performance throughout enabling it to make confident strides to ensure long term success!

III. NEED FOR ACHIEVING COMPETITIVE SUPERIORITY

Need for Benchmarking

An enlightened manager of the aspiring firm, should make every possible endeavour to introduce *benchmarking concepts* irrespective of the scale of operations. Even small and medium size companies, should begin adopting *benchmarking processes* analogous to those of larger corporations in the interest of ensuring the future prosperity of the organisations. Bench marking would certainly stimulate the modern firm to borrow and adopt the *best way of doing a task*, which is found satisfactory by one and all.

Since competitive edge depends overwhelmingly on the type of technology being used by any modern firm, it should be able to deal effectively with business co-partnerships, focus on core competencies and outsourcing of certain activities from the point of view of overall healthy growth. The strategies to be adopted in the context of accelerating pace of changes taking place should thoroughly be grasped, keeping in view the changing needs of customers, markets, technologies and product innovations.

The foremost challenge would hinge on the right kind of leadership. Organisational operations will have to be undertaken only by a CEO having the 'global mind set' in order to make the modern enterprise a thriving concern. The greatest challenge is to put on the market, a product of excellent quality at a price lower than the one charged by its competitors.

Since successful implementation of 'bench marking practices' presents unique problems, there is need for selectivity in implementing improvements in important processes, after carefully cross checking the positive benefits, and possible improvements in performance standards. The practices adopted by leading and reputed organisations could be studied at the outset, before embarking upon a suitable plan of action.

The widely acclaimed *Malcolm Baldridge National Quality Award* is a top award which had emanated from the USA and it is conferred on a top class firm meticulously assessed for excellence in quality on the basis of 7 core values being bench marked as outlined below:

Table 1.1: Malcolm Baldridge National Quality Award

S.No.	MBNQA Core Values	Points
1	Leadership	90
2	Information Analysis	80
3	Strategic Quality Planning	60
4	Human Resource Development	150
5	Process Quality Management	140
6	Quality and Operational Results	180
7	Customer Focus and Satisfaction	300
	TOTAL	**1000**

Other Important Factors

Organisational improvements would be the most effective only in an environment which has the inherent drive to innovate. The aspiring firm should have certain important objectives while undertaking bench marking practices.

(i) To increase the market share substantially
(ii) Cutting down operating expenses
(iii) Exploring new markets

The ambitious firm, besides having the objective of improving the current level of performance, should aim at exceeding another leading competitor firm's performance or bench mark !

Since in today's environment even the very survival in the market requires innovation the modern firm should have several *varied dreams of success* and try to fulfil these by a judicious combination of 'qualitative' and 'quantitative' measures, at every stage of production and distribution process at the earliest opportunity.

Besides, the CEO/Manager must endeavour to live in the present scenario with the propensity to learn from the *experience gained in the past*. He should constantly keep in mind the healthy growth of the firm to reach the 'long term goals', 'objectives' formulated.

In the interest of ensuring *healthy growth*, activities should be well structured over a timeframe and there should be an overall rhythm in the functioning of the firm, in spite of periodic changes taking place in the external environment.

In addition, the modern firm should make every possible endeavour to learn from failures and undertake effective probe for reaping both 'short term' as well as 'long term' benefits.

Flexibility in Approach and Intangible Benefits

In a dynamic scenario, ultimately flexibility alone could provide the necessary foundation for achieving *competitive advantage*. (See Figure1.1)

Figure 1.1: Flexibility and Competitive Advantage

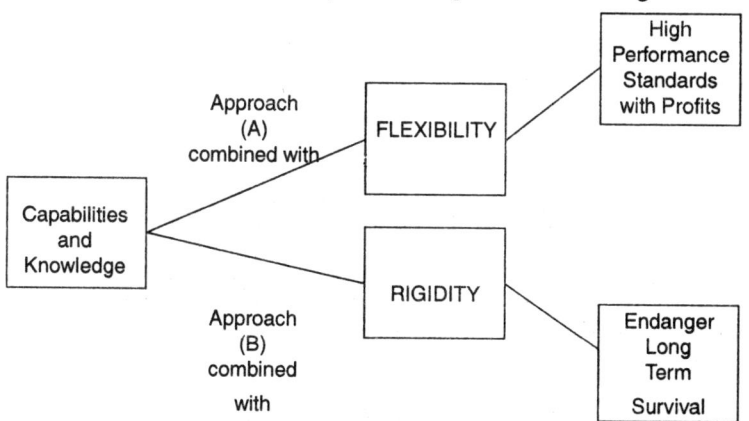

As could be seen from the figure 1.1, Approach (A) would lead the modern firm towards acquiring *sustained competitive advantage* and help to reap greater profits.

Whereas, Approach (B) would certainly endanger the healthy growth of the firm in the long run.

Besides, *'knowledge and capability'* indeed are more powerful tools than other resources and only the organisation, which could wield these tools effectively, would emerge successful in the long run.

Firms such as Toyota, Microsoft and Wall Mart surpassed their rivals – General Motors, IBM and Sears only on account of the fact that they identified and deployed *intellectual capital*, viz., the *collective brain power* in order to enhance *efficiency* substantially, and reach greater heights of excellence.

Besides, there are certain other *intangible assets* on which the modern firm could lay accent, so as to facilitate *'value creation'*.
  (i) High level of customer service and satisfaction
  (ii) Harmonious relationships with suppliers
  (iii) Leadership qualities which help to catapult the firm to the top league.

2

Strategic Thinking: Key to Global Competitiveness

> *"Should you find yourself in a chronically leaking boat-energy devoted to changing vessels is likely to be more productive than energy devoted to patching leaks"*
>
> —Warren Buffet

Formulation of a suitable strategy to achieve success is a highly fascinating process. In the turbulent 21st century environment, the road map which a modern firm should traverse has to be meticulously planned, keeping in view the need for 'value building offerings' with an accent on having always an edge over its competitors.

The importance of competitiveness has been receiving greater attention in recent years. The socio-economic health of a country is greatly dependent on rapid industrial development with multifarious product lines comprising healthy and highly competitive firms.

Poor competitiveness has been identified as the root cause of the numerous problems being confronted by developing countries all over the world.

In the highly competitive 21st century milieu, the very survival of the firm or industry depends mostly on the ability to turn out quality products and this could be made possible only by having a quality conscious work force.

The foremost requirements of a global player could be broadly classified as outlined below:
- Producing the right product
- Producing at the right time
- Producing at the right place
- Ensuring consumer satisfaction and endeavouring to exceed consumer expectations.

The exposure to rapid globalisation invariably involves abandoning outmoded techniques and primitive practices, and adoption and enhancement of sophisticated practices which would conform to international standards.

At present, the environment is rapidly changing, and the modern manager would certainly perceive that the life span of opportunities is dramatically decreasing.

To achieve sustainable competitive advantage, choosing an optimal positioning within the market alone would not suffice. In order to be really successful, there is need for reorienting the internal structure with greater accent on 'core competencies and capabilities' which would ultimately pave the way for rapid expansion and further development.

In this context, the role of the CEO assumes paramount importance. Depending upon the scenario which materialises, the top management should be in a position to make quick manoeuvres. In this regard, it may be necessary to invest in ways by which new competencies could be suitably developed.

The modern firm can carve out a 'suitable strategic direction' only after considerable experimentation in regard to the emergence of new markets in the current scenario. Here, the firm should adopt a selective approach, so that the investments prove highly fruitful. The firm should endeavour to incorporate adequate flexibility in its strategy concerning the decision making process.

In this regard, the modern firm should:

(i) Explore new opportunities in consonance with its approach in developing competencies and capabilities.

(ii) In a highly volatile business environment, the strategy should develop a suitable platform that gives adequate access to future growth opportunities-flexibility in approach is much required.

(iii) The platform constitutes the base for potential opportunities, and it should be quite broader in range in order to be really valuable.

(iv) Only with the flexible production system, there can be enough scope for switching over from one product type to another. Good platform decisions combined with the objective of 'sustainable value creation' for the firm should be accorded priority.

The modern firm, keen on having a 'global strategic orientation', invariably makes no distinction between domestic and foreign market opportunities. By and large, it endeavours to serve an essentially identical market appearing in several countries around the world. It concentrates mainly on developing suitable global strategies to compete with other global firms.

Though initially a modern firm may be highly supportive of its short term strategic objectives and goals, its foremost long term strategy would be to reach higher levels of development, which would facilitate its potential contribution to substantial globalisation benefits.

Dr. Levitt had suggested that in order to be truly competitive in the world market firms should shift their emphasis from local 'customised products' to 'globally standardised products', that are advanced, functional, reliable and low priced.

Prof. Buzzel has outlined the benefits emanating from product standardisation as indicated below:

- Economies of scale

- Faster accumulation of learning experience
- Reduced costs of design-modification

Thus, it may be concluded that most of the Japanese and European firms which offer standardised products have higher levels of 'product and process innovations' and thereby acquire competitive advantage over those firms which mostly rely on 'product adaptation'.

Effective Functioning of Modern Firms

In the turbulent competitive environment, the foremost function of a modern firm would be the job of creating value which would constitute the first step, and would also ensure superior performance. Any modern organisation can hope to achieve superior performance only by doing something in a unique way that would be most difficult for other firms to emulate.

In a highly competitive environment, the consumer has several alternatives and ultimately the success of the modern firm depends on its capability to make strategic choices by determining the products which could be offered with great advantages to the firm, and which would also offer greater appeal to the existing consumers.

The flow of latest information to the modern firm should be on a solid-basis, and with timely information and using great sagacity to maintain the requisite inventory, it can forge ahead rapidly.

Strategy is concerned mostly about winning the battle by intensely focusing on the style and fashion sought by the consumers and thereby creating 'value'.

The Inspiring Story Of Eastman – Kodak

In 1994 George Fisher, who had an excellent track record at Motorola, assumed charge as CEO of Eastman Kodak. He formulated Kodak's 'high tech growth strategy' and churned out the most impressive array of digital cameras, scanners and so on at phenomenal speed. Having an excellent product breadth, Eastman Kodak, under the eminent leadership of Fisher surpassed several renowned competitors by investing more than $ 450 million every year into research and product development.

Fortunately every year the sales have started increasing at 25% and Fisher is endeavouring his best to achieve technological leadership by throwing overboard the old, slow bureaucratic culture. He has been able to put on the market, new state of the art digital imaging products by adopting an innovative 'pay for performance' system. He works with the firm conviction that winning the technological arena ultimately depends upon speed and flexibility.

There is a never ending race between Eastman Kodak and other renowned firms such as Hewlett Packard, Fuji, Canon and so on, in respect of offering innovative products in digital technology, keeping pace with latest developments in the field.

Importance of Planning

Indeed, Eastman Kodak had formulated mind-boggling plans for its healthy growth on the basis of inspiring leadership provided by George Fisher.

The importance of formal planning, with systematic decisions about goals and activities to be pursued by organisations, has grown dramatically. Ambitious and aggressive entrepreneurs keen on utilising the latest opportunities invariably engage in formal planning.

The following figure 2.1 provides an outline of the basic decision making steps, which an aspiring firm usually undertakes in a formal planning process.

Strategic Thinking: Key to Global Competitiveness

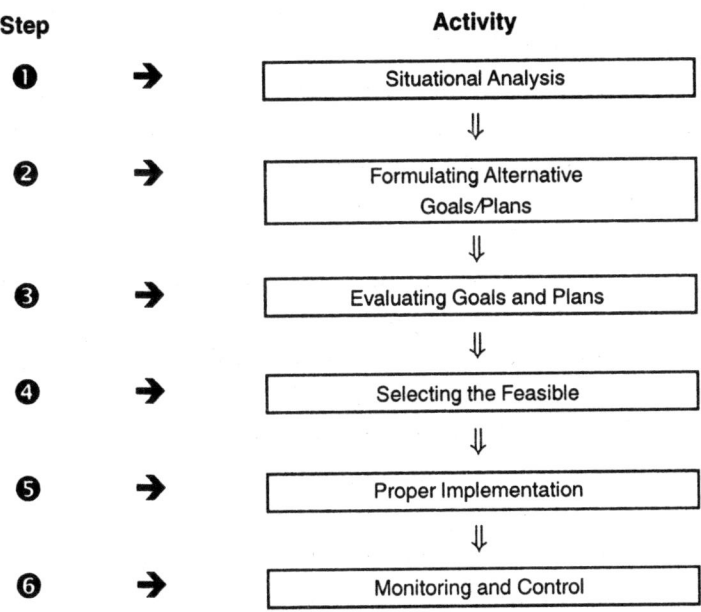

Figure 2.1: Formal Steps Involved

The modern firm takes the first step by undertaking situational analysis in a thorough fashion. Current conditions are fully analysed and diagnosed, so as to make forecasts of future trends. The management gets a bird's eye view of several alternative plans to achieve specific goals and this requires a lot of creativity/participation of employees. As such goals are specific targets which the management endeavours to reach in a specific time frame and these goals should certainly be challenging as well as realistic.

Jack Welch of General Electric always focused attention on world class customer service and formulated plans with the avowed objective of excelling all the rival firms. Decision makers of modern firms prioritise their goals and eliminate the unimportant. During evaluation of major planning efforts, new ideas are constantly reviewed and existing approaches are refined.

Keeping in view the predicted set of circumstances, goals and plans which are most appropriate and feasible are taken up allowing much flexibility.

After finalising the best plans and goals, the management enters the implementation phase. The successful implementation of the plan depends on the following factors:

(i) budgetary system with sufficient financial resources to execute the plans formulated.

(ii) Suitable reward system to facilitate rapid implementation with incentive programmes—higher salaries, bonuses, promotions, etc., so that employees remain always motivated and committed.

Above all, the final step relating to the planning process would be concerned with monitoring and controlling the actual performance. The top level managers should not hesitate to take corrective action depending on the changing situations as well as improper implementation, they notice periodically.

The success of the modern organisation depends upon certain critical requirements and these factors are linked with the 'strategic planning process'.

As such 'strategic management' involves managers from all parts of the growing organisation.

Strategic planning activity could be broadly classified into:

(i) Focusing on long term externally oriented issues for the benefit of the entire organisation.

(ii) Short-term, tactical and operational issues with specific budgets and staff units.

In a modern firm the CEO, as well as a group of top managers mostly furnish the strategic direction or vision for the organisation. The remaining tactical and operational managers regularly provide valuable inputs so as to achieve outstanding success.

As such there are six vital factors connected with achieving excellence in any management process:
1. Formulation of organisation's mission, vision and goals.
2. In-depth analysis of changing environmental opportunities and threats.
3. Complete analysis of internal strengths and weaknesses.
4. Undertaking SWOT Analysis
5. Implementation of strategy
6. Suitable strategic control system.

1. Mission, Vision and Goals

The following figure 2.2 provides a bird's eye view of the spectrum of activity concerning the strategic management process.

Figure 2.2: Modern Firms-Strategic Management

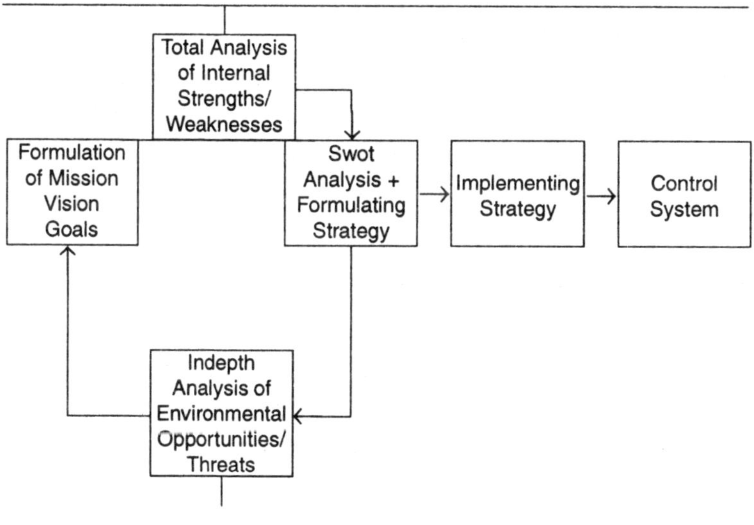

```
                  Total Analysis
                   of Internal
                   Strengths/
                   Weaknesses
                        │
                        ▼
  Formulation         Swot
  of Mission        Analysis +      Implementing      Control
    Vision         Formulating  →     Strategy    →   System
    Goals           Strategy
      ▲
      │             Indepth
      │            Analysis of
      └──────── Environmental ◄──┘
                Opportunities/
                   Threats
```

In order to be successful every business unit requires a direction, which is mostly influenced by its 'mission statement'. Its 'corporate vision' would be highlighted only by analysing the mission statement, which presents the organisation's overall philosophy. The key-traits of the organisation's beliefs and its expectations from its employees must be well articulated.

A modern firm striving for healthy growth could be humane as well as profitable. Besides, it should develop an obsession for values-which would help to build a 'value culture'.

Most modern firms, which remain in the forefront/and continue as winners, are the ones who commit themselves ardently to the value path and stay on it no matter how much it hurts. It is built on solid foundations to equip them to survive the tough times and exploit the good times.

2. In-depth Analysis: Environmental Opportunities

a. Analysis of Industry/Market

There is need to identify major product lines, and critical market segments. Besides, growth rates applicable to the entire industry/key market segments will have to be ascertained.

b. Competitor's Activities

Besides major competitors in the product line, and the market shares enjoyed by them, the strengths and weaknesses of different competitors, uniqueness of different products offered by them, and their position in regard to cost-leadership enjoyed will have to be thoroughly analysed.

c. Regulatory Issues: Political/Social Milieu

How the existing legislation and regulatory activities of the government will exercise influence on the growth of the industry-associations, influencing the productive activity in the specific product line should be thoroughly assessed.

d. Human Resources

Availability of skills, needs, shortages, opportunities and key labour issues will have to be examined.

e. Macro-economic Picture

Important economic factors which mostly affect supply, demand, growth rates, competition and profitability should be clearly understood.

f. Technological Factors

The industry would be much affected by recent scientific breakthroughs, changes in technology and innovative methods. The aspiring entrepreneur must be constantly on the alert to implement necessary modifications.

An opportunity staircase for the modern firm could be worked out on the following lines:

Figure 2.3: Opportunity Staircase for the Modern Firm

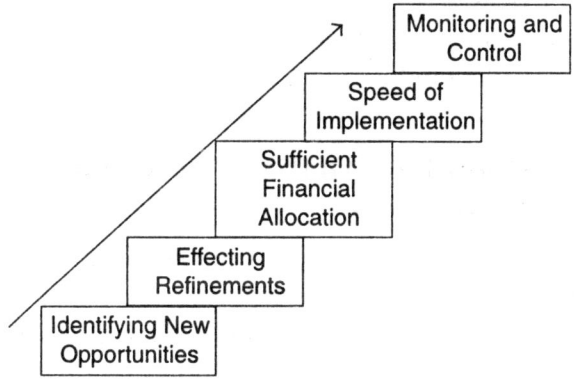

When the entrepreneur discovers or identifies a new opportunity that would bring about diversification and help the modern firm to forge ahead with substantial profits, it should make every possible endeavour to manufacture and market the product at the earliest in order to seize the new opportunity.

After ascertaining the psychological factors affecting the potential consumers in the current milieu, it should initiate the project to produce a specific new product with due regard to existing capabilities, new technologies required and so on.

It would be worthwhile having greater interaction with potential customers prior to the taking up of its manufacture and supply, since it would provide deeper understanding of the market trends, and the new opportunity presented to the entrepreneur could be advantageously capitalised.

In this regard speed, accuracy and timing are the essential ingredients which must be given particular importance for the successful implementation of the strategy.

The firm should be able to allocate substantial funds which would facilitate the successful fruition of the project in a brief span of time. The firm should not hesitate or get bogged down in bureaucratic controls, and cumbersome procedures as regards the allocation of adequate funds.

Besides having a feed back mechanism on the consumers' viewpoints and achievements in the sales front, the firm should tailor the product in an unique fashion so as to captivate the consumers on a massive scale.

3. Complete Analysis: Internal Strengths and Weaknesses

Internal analysis helps strategic decision matters of the modern firm in the following manner:

   (i) It provides an inventory of the firm's multifarious skills and resources.
   (ii) Better insight into its overall as well as its functional performance levels.

As such the resources of a firm could be termed as inputs which could be accumulated overtime to enhance and refine the performance of the organisation.

Besides, resources cover factors such as:
   (i) Financial strengths/weaknesses of the firm
   (ii) Strengths/weaknesses at different levels of management, key human resource activities, training programmes, labour relationships, incentives, promotions, etc.

(iii) Key market segments explored and the competitive position enjoyed.
(iv) Strengths and weaknesses of manufacturing activities, production service/delivery arrangements.
(v) Research and Development activities which keep pace with latest developments.
Resources could be classified into:
(a) Tangible assets
(b) Intangible assets

Tangible Assets

These assets cover production equipment facilities and raw materials used.

Intangible Assets

These would encompass the reputation of the firm, image projected, cultural factors, technical excellence, patents significant for the firm, etc.

Resources invariably constitute the fulcrum for the competitive advantages which a firm could very well enjoy in relation to the rival firms.

(i) The resources should be instrumental in creating 'customer value' which is unique.
(ii) For reputed firms such as Dupont, Merck, etc., certain patented formulas which the companies exclusively employ for surpassing competitors would constitute the most highly valuable resources.
(iii) To achieve 'customer driven operational strategy', leading firms take up manufacturing-resource planning, as well as distribution-resource planning, and thereby ensure 'total quality management.

Core Competencies

Well-organised, rare and inimitable resources of a highly successful firm may very well be termed as its 'core competencies'.

Such core competencies refer to the unique skills as well as technical knowledge an organisation possesses and these core competencies provide the required edge to surpass its competitors.

Benchmarking Practice

Benchmarking is the most efficient approach to promote effective change. Prof. Michael Spendolini has aptly termed benchmarking as "a continuous systematic process for evaluating the products, services or work processes of organisations, that are recognised as representing the 'best practices' for the purposes of organisational improvement."

As such 'benchmarking' is a measurement standard or reference utilised for comparison and it is the continuous process of identifying, adapting and understanding the best practices and processes that would facilitate the firm to achieve superior performance. It measures an organisation's products, services as well as processes with meticulous care in order to establish targets, priorities and worthwhile improvements which in turn would lead to:

- competitive advantage and
- reduction in operating expenses

Benchmarking promotes effective change in an organisation by learning from the successful experiences of other firms and by putting that learning to stay ahead and maximise benefits.

The benefits resulting from benchmarking practice could be broadly enumerated as given below:

(i) The organisation as a whole gets a better understanding of the current situation
(ii) It encourages innovation
(iii) It heightens sensitivity and adaptability and
(iv) It helps to establish appropriate stretch goals, and implement realistic action plans so as to enter new markets and increase market share.

Strategic Thinking: Key to Global Competitiveness 31

The concept of benchmarking is reported to have originated from Japan especially after Rank Xerox Corporation, established a novel programme to study its key processes against world class companies.

Benchmarking programmes initiated by 'Xerox' brought about remarkable changes in eliminating inefficiencies and improving competitiveness. Several other companies such as Hewlett Packard, Ford, Corning etc., also initiated such programmes and made rapid strides in a short span of time.

In recent years, M/s. Siemens the renowned German firm, producing sophisticated electrical and electronic products/gadgets has benchmarked itself quite extensively in several areas, including customer service, with a view to tackle all its rivals, and has also achieved recognition as 'being the best of the best'.

4. Understanding SWOT Analysis

Strategic decision makers in an organisation could get the required information to formulate suitable business/functional strategies after analysing the external environment as well as the internal resources.

A detailed comparison of strengths, weaknesses, opportunities and threats would be normally referred to as a SWOT analysis.

By adopting SWOT analysis:
- *(i)* A firm could identify and utilise its strengths in an advantageous manner and could also *capitalise on opportunities.*
- *(ii)* It could deftly counteract threats being posed and alleviate internal weaknesses.
- *(iii)* While formulating a suitable strategy, instead of analysing the present position, the firm could devise a coherent course of action to reap rich dividends.

The foremost aim of the modern firm is to reach enhanced competitiveness. As such the 'SWOT analysis' provides an wonderfully fine overview of the firm's business position, and indicates whether it is fundamentally healthy or unhealthy.

(A) Identifying the Firm's Strengths as well as Resource Capabilities

'Strength' may assume several forms:

(i) *Skills/Expertise:* Impressive track record in defect-free manufacture, low cost know-how, excellent customer service, capability to produce suitable innovative products and impressive promotional efforts undertaken by the firm.

(ii) *Physical Assets:* Assets which are highly valuable like state of the art plant and equipment, excellent distribution network, selective locational factors, access to natural resources and so on.

(iii) *Human Factor:* Highly capable and well-experienced managers, workers and talented personnel who provide worthwhile contributions on a continual basis.

(iv) *Organisational Factors:* Total quality management – systems and top commitment, customer satisfaction as the core strategy with a large base of loyal customers.

To enhance organisational performance, suitable *informational flows* from sources outside the organisation should be ensured. Besides, a strong balance sheet and credit rating with a fine financial snapshot are the essential attributes which speak for the strengths of a firm.

(v) *Intangible Assets:* Capability of the brand to communicate a rich set of images would certainly be a great asset. The reputation enjoyed by the firm, the good will it has earned among customers, the high degree of employee – commitment, a sense of identity among them, an organisational culture and ambience that would facilitate the optimisation of the potential of all the employees are indeed great assets.

(vi) Possessing Competitive Edge: Short development periods to put new products on the market, highly flexible manufacturing capability, a marvellous dealer network, an excellent R and D unit, which would constantly accentuate efforts to bring out new products after monitoring consumer tastes may facilitate the firm's rapid progress. Besides, the innovation process could be speeded up by drawing the best talents within the organisational set up.

The firm's success depends on seeking continuous improvements by using its distinctive competencies in specialised areas as well as rewarding and preserving distinguished craftsmen; organisations become 'distinguished' because they master multifarious details.

(B) Identifying Weaknesses/Deficiencies

As regards the firm, a weakness could be termed as a condition, which puts it in a clearly disadvantageous and hence vulnerable position in relation to others. Mostly the internal weaknesses of a firm could relate to:

(i) Deficiencies in essential skills or expertise – which would help to achieve 'competitive edge'
(ii) Lack of human, physical and organisational assets inclusive of intangible assets necessary for competition.
(iii) Disadvantageous position in competitive areas.
(iv) Lack of world-class R and D facilities and personnel.
(v) Too narrow a product line due to absence of diversification.
(vi) Existence of under-utilised plant capacity.
(vii) Higher overall unit costs in relation to those of its foremost competitors.

(viii) A weak balance sheet with considerable debt burdens.
(ix) A weak brand image lacking in emotional uniqueness.
(x) Imperfect product in terms of quality.

(C) Identifying Potential Opportunities

Core Competence

A 'core competence' is considered central to a company's competitiveness and profitability. It would relate to a firm's scope and depth of technological know-how or to a combination of special skills which are highly valuable to make it competitive.

Normally core competencies reside in a company people- staffers and workers and not in other factors.

There are several core competencies which a firm can advantageously utilise:

(i) Special skills for manufacturing high quality products
(ii) Suitable know-how and an operating system for effectively tackling customers' orders swiftly and accurately.
(iii) Speedy implementation of innovative ideas and putting new products after thoroughly researching *customer needs, tastes* and *latest fashion trends.*
(iv) Providing excellent after sales service/arranging for impressive product display and selecting ideal retail locations will confer innumerable benefits.
(v) A core competence constitutes a basis for competitive advantage only when it is a distinctive competence i.e., capability to perform well with an edge over rival firms. Motorola's distinctive competence is largely attributed to its virtually defect free manufacture owing to the adoption of 'six sigma quality standards' (an error rate of one per million pieces). Thus it has achieved world market leadership as regards cellular telephones.
(vi) Firms are not alike, as regards resources such as skills, and various types of assets – covering physical, human,

organisational and intangible. Besides, competitive capabilities and market achievements differ between firms. As such different companies possess varying resource strengths and weaknesses.

Therefore, a modern firm can hope to achieve outstanding success only when it has appropriate and ample resources with which to compete with rival firms. It derives competitive vitality and achieves sustainable competitive advantage by having a combination of good and substantial resources.

Thereby it would be in a position to craft or tailor an attractive strategy by suitably leveraging its resource capabilities.

By always focusing attention on building a strong resource base for the future, the firm would be in a position to 'maintain competitive superiority'.

Market Opportunities

A wise market strategist would be always alert to the opportunities thrown open to the firm. However, the decision to take up the manufacture of a new product depends to a large extent on its resource capabilities. If the requisite resource capabilities are missing, the firm will have to take aggressive steps to develop or acquire the missing resource capabilities, provided it could hope to outperform the rival firms.

The methodology of approach adopted by Boston Consulting Group USA could be taken up with advantage by any aspiring firm, while formulating the marketing strategy.

In Figure 2.4 presented hereunder, 'product strategy' is assessed on the basis of a box type analytical tool.

Box (A) The business having healthy growth trends, reflects good growth rate as well as a high market share. The strategist may have to decide whether augmenting his investment would help to increase the market share enjoyed by the firm.

Figure 2.4: Product Strategy-Analysis

(A)	(B)
Products falling in this sector have 'High Growth Rate' as well as 'High Market Share'	Products classified in this sector may be having 'High Growth Rate', with possibilities of 'Low Market Share'.
Generates substantial 'Cash' but requires injection of 'Sizeable Cash' at the outset	Enhancement of 'Market Share' with additional injection of 'Cash' possible
(D)	**(C)**
Products in this category enjoy 'High Market Share' with Low Rates Of Growth.	Products classified in this Group have Low Growth Rate as well as 'Low Market Share'.
These products generate fairly substantial Cash Flow	The Firm can persevere with such products provided – 'New Improvements' could improve business.

High ← Market Share → Low

(Rate of Market Growth (Annual): High at top)

Box (B) This category is known as 'star quality segment' where it is essential to boost investment in order to sustain the star quality enjoyed.

Box (C) Indicative of high market share, it generates fairly substantial cash and is known as 'cash cow' segment. Since the growth rate is somewhat low, the firm need not invest too much in this sector.

Box (D) Here business is at a low ebb. In case, it is not possible to augment earnings by effecting improvements, it has to be abandoned. If new improvements are helpful, the opportunity could be pursued.

(D) Identifying the Threats Posed

The following factors could be the possible threats to the market position enjoyed by the firm.

(i) Emergence of cheaper technologies and their adoption by other firms.

(ii) The rival firms may introduce innovative and superior products.
(iii) Possibility of entry of lower cost foreign competitors into the firm's market arena.
(iv) Cumbersome new regulations which may affect the firm's activities/projects.
(v) Adverse changes in foreign exchange rates as well as political upheavals in a foreign country where the company has a strong base.
(vi) Shifts in consumers' needs and tastes quite away resulting in lower demand for the firm's products.

Ultimately, the success of the firm's strategic activities are greatly dependent on its 'future profitability' as well as the market position it hopes to enjoy over the years.

The aspiring firm, therefore, resorts to every possible endeavour that would help the firm in 'tailoring the most appropriate strategy' that would confer the greatest benefits which entails-

(i) Pursuing specific market opportunities well-suited to the firm's resource capabilities.
(ii) Constructing a resource base that would defend the firm from all external threats.

Thus, the meticulous adoption of 'SWOT' analysis facilitates worthwhile decision making by the modern firm in a turbulent environment, with due regard to its strengths, weaknesses, opportunities and threats.

(E) Implementation of Strategy: Achieving Cost Competitiveness

By examining the make up of a firm's own 'value chain', and comparing it to that of the rival firms, it would be possible to indicate which one has much of a comparative cost-advantage or disadvantage. Different cost components responsible for the

situation could also be ascertained. This information is of vital importance while planning a suitable strategy for future growth.
The following strategic steps would be found useful.
   (i) Negotiating more favourable prices with suppliers
  (ii) Cutting delay in paper work proccdures
 (iii) Avoiding bottlenecks and inefficient work flows or procurement procedures
 (iv) Re-engineering different business processes by:
     (a) Boosting employee productivity
     (b) Improving efficiency of key activities.
  (v) When products, systems or services are designed, at every stage the costs should be properly evaluated. Value analysis techniques could also be adopted; possibilities for simplification could be worked out.
 (vi) Suitable investments in cost saving technological devices, flexible manufacturing techniques, automation etc., could be considered. Ultimately, the modern firm should make concerted endeavours towards:
     • Reduction of scrap
     • Reduction in total cost of quality
     • Reduction in manufacturing lead time

(F) Suitable Strategic Control System

In the 21st century milieu, decision making becomes a highly critical and complex task. On the one hand, economic competition will reach unprecedented heights; and on the other, amidst explosion of knowledge managers should possess great self-confidence and business acumen in exploiting wonderful opportunities being presented.

The modern firm should have an appropriate strategic control system with capability to utilise 'the information flows' most effectively and endeavour to develop successful new products. Besides, in order to function effectively the following steps are quite essential.

(i) Chief Executive Officers/Managers keen on launching new products, and effecting improvements should resort to 'total quality management programmes' inclusive of improvement of the bottomline. The foremost objective would be higher profits, reduced costs, and more satisfied customers.

(ii) It has to be ensured that officers in senior management level are highly committed. This requires involvement and taking personal responsibility for achieving success in the field of specialisation.

(iii) Nowadays several leading organisations approach the task of a 'Baldridge assessment' by entrusting the 'seven categories' among seven team leaders.

(iv) Recent developments in communication technologies covering tele conferencing, video taping and photocopying could be instrumental in improving communication in large-scale assessment programmes.

(v) The modern organisation should formulate a common training strategy for team leaders, who in turn would train the team members. The objective is to achieve consistency in the way different team members function.

(vi) Sometimes, there are possibilities of serious challenges being faced by employees without perceptible quality improvements in sight. Periodic assessments and efforts to *buy in across the organisation* would help to ameliorate such problems. The spirit of continuous improvement has to be inculcated inspite of setbacks and delays.

To Sum Up

The modern firm aspiring for *world class standards* should be able to craft suitable objectives and draw up plans covering all aspects of manufacturing as indicted below:

(i) Excellent product design
(ii) Suitable equipment and process

(iii) People and their training requirements
(iv) Inducting flexibility and reducing all rigidities
(v) Manufacturer vs. buyer decisions
(vi) Appropriate systems and procedures for planning, controlling and monitoring.
(vii) Selecting suitable suppliers
(viii) Adherence to high quality performance in all areas

Professor David J. Collins in his publication *Creating Value: Successful Business Strategies*, has presented a remarkable exposition on the framework of 'business strategy'.

He has categorically stated that when a firm offers a 'single product' as competitive offering or a 'cluster of products' as part of corporate strategy, the foremost principle has inherently a 'financial goal' i.e. making the firm more valuable to its owners in the 'long run'. This principle has universal applicability since it is equally relevant for a tiny unit as well as a giant corporation. If the business finds it difficult to earn at least its cost of capital, certainly it can not survive independently.

Competitive strategy is concerned with advantageous 'value building offerings'. As such, an offering is expected to build value to earn more than its cost of capital.

When a favourable market opportunity is presented, a firm is concerned more with the *financial objective,* and tries to assess whether its distinctive resources would provide an edge over its competitors.

The condition could be well-presented with an analogy of a pair of scissors with (i) attractive market position as a blade and (ii) the key resources readily available as the other blade. Here, two kinds of strategy overlap like the two blades of a pair of scissors which can only cut together in order to turnout 'value adding offerings' in the form of products found useful by the present day consumers.

Figure 2.5 presents how the two strategies overlap and function.

Figure 2.5: Pair of Scissors Analogy

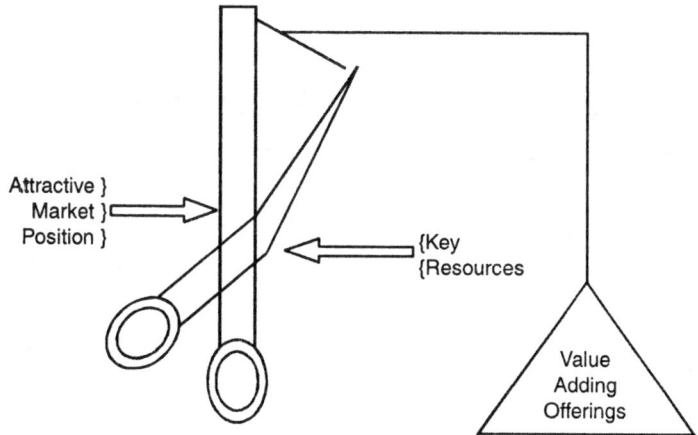

Only a healthy firm having the required foresight with winning resources would emerge successful in this context.

3

Technological Competence

"The guts to attack yourself mercilessly is what counts!"
—Tom Peters

In the 21st century milieu, technology has to be managed most effectively, for the generation of wealth. Developing nations have realised the fact that only by 'technological strength' they can hope to achieve economic growth. Nowadays there is greater awareness among entrepreneurs of small countries and firms adopting obsolete technologies and obsolescent production facilities that they will not at all be able to survive and grow, in the present highly competitive global scenario.

It is an acknowledged fact that competitive-success goes increasingly to dynamic enterprises which can absorb, apply and co-ordinate new technological developments very rapidly. To meet the exclusive needs of customers, many successful firms across the globe adopt several tools and techniques to achieve commercialisation of several innovative ideas. To forge ahead in a specific product line, the aspiring firm, makes every possible endeavour to effect continuous, incremental improvements in 'function, cost as well as quality'. Formulation of cost effective designs, new process/product technologies and flexible manufacturing systems are increasingly being adopted.

On Technology

The term 'technology', known as the engine of change, encompasses diverse collection of tools, instruments, machines, new techniques organisational methods, systems and so on.

In the contemporary scenario, it has a wide variety, besides having much ambiguity.

Technical innovation could be classified into three distinct phases:

(i) The first phase is concerned with invention where a scientist or a technically qualified person puts forth a creative idea.

(ii) The second phase pertains to actual exploitation wherein practical-methodologies/applications are fully developed.

(iii) The third phase is more concerned with diffusion of technology where other firms/people perceive the distinct advantages of the new approach and begin applying it.

As such, the process is self-stimulating and self-perpetuating, owing to the realisation that other firms/manufacturers appreciate the distinct advantages of the new approach, and in order to survive in a highly competitive milieu begin to adopt the new approach and also endeavour to improve the same.

Mainly with a view to gain competitive advantage, firms constantly resort to innovations in respect of developing new products, new production processes and new services. As such change has become an important prerequisite for achieving excellence in today's life style.

In the beginning of the 20th century, mass production techniques assumed great significance in view of the constantly expanding market opportunities. The credit for introducing assembly line production goes to Henry Ford whose experimentation in automobile production, yielded spectacular results. Adam Smith, the renowned economist, introduced the "concept of interchangeability" to augment the rate of assembling.

It may be stated that assembly line costs and time span were considerably reduced subsequently; after Sloan's behavioural study findings were supplemented to the 'moving assembly line' concept introduced by Henry Ford.

The global socio-political framework had undergone a rapid change, soon after the World War II. Technology transfers between countries became the order of the day, and several nations vied with each other to capture substantial market shares in specific product lines, in which they were dominating.

It is an acknowledged fact that life in the developed world would be completely transformed, on account of spectacular advances in technology having practical commercial application.

The following table would furnish the advances effected in recent decades in different areas:

Table 3.1: Advances Effected in Recent Decades

S. No.	Area	Advances/Changes
1	Housing-Modern	Little progress has been made. Modern homes are homes more practical and little more comfortable.
2	Air Travel	Incremental advances; real cost of flying across Atlantic has been reduced.
3	Motor Cars	Incremental advances; comfort and performance of the 2000 small car is similar to the luxury car of 1960. Fuel efficiency has been achieved; several models are available with wider choice.
4	Consumer Durables Refrigerators Audio Systems Colour Televisions Motor Cycles Sewing Machines Mixers/Grinders	Made available for wealthy homes in 1960's... Now even middle income families can afford.
5	Telephone Service/ Trans-Atlantic Calls	Considerable reduction in rates/costs

Table 3.1 contd...

Technological Competence

Table 3.1 contd...

S. No.	Area	Advances/Changes
6	Personal Computers	In 1960: were bulky expensive After 1990: substantial no. of households own PCs; Prices affordable.
7	Food Retailing	Unexpected rapid advances have taken place. Highly efficient service combined with high responsiveness to consumer preferences/tastes.
8	Video Camera	Now widely utilised as tool for security reasons also.

Besides futurologists, scientists and technicians have predicted certain important changes on account of technological advances:

From the point of view of better management, entrepreneurs of the future would prefer small plants since it would be easier to motivate people, and generate the required team spirit quite easily to augment productivity.

Besides, strikes would be less frequent with better labour relations. The 'timespan' for managing the plant could be considerably reduced.

In mature industrial economies, entrepreneurs would prefer small industrial manufacturing units (plants) providing greater 'flexibility' of production. As a result several batches of new products could be taken up for purposes of testing market acceptability without disruption in normal production activities. Besides, entrepreneurs need not spend vast sums of money for installing new plants; instead they can install small plants with little gestation periods.

The manufacturing units of the future would be able to turnout batches of individual products meticulously crafted to meet the fashions required in a particular market segment instead of producing long lines of identical products.

Modern consumers exhibit considerable aversion for product churning once adopted by the Japanese, for which there was considerable consumer revolt. Nowadays consumers want exceptionally fine products to suit their tastes, and not products with minor cosmetic variations. In the circumstances, it is not possible to predict how consumer tastes are likely to undergo change over the years. Only installation of small plants replacing the conventional factory system could solve future problems, with greater flexibility.

New Technologies by 2020

The practical possibilities of amazing new technologies gaining popularity by 2020 are outlined below:

    (i) The world's first superfast *Magnetic Levitation (MAGLEV)* passenger train began running on 1st January 2003 from Shanghai in China. It was able to reach 500 km/hour, and it had covered 30 km of distance within 7 minutes duration. It was built by Transrapid and Co. of Berlin in Germany.

    Such trains are becoming popular in England, Japan, Germany and so on, and by 2020, several other countries may put into commercial use such 'Maglev' trains for the benefit of passengers.

    (ii) Another prominent technology gaining much popularity is *carbon fibre technology*. Several firms had already started manufacturing *sports equipment* such as tennis racquets, skis etc. In the global platform carbon fibre technology is finding greater application on account of dynamic improvements in efficiency noticed by transport professionals, and already experimentations are being undertaken for the enhanced use of this technology in aircrafts, passenger cars, electric cars and so on.

Lasers: Offering Vast Potential

The availability of high power lasers have stimulated much interest in laser manufacturing and it holds vast potential offering several unique advantages.

Some laser based operations have already assumed greater popularity in manufacturing shop floors of several modern firms of the globe as outlined below:

   (i) Laser cutting   For precision drilling
  (ii) Laser welding   Possibility of welding dissimilar materials
 (iii) Laser heat treatment
 (iv) Rapid proto typing, etc.

Besides, lasers find application in surface alloying, cladding, glazing and so on. Several new materials possessing novel properties could also be turned out by the use of lasers.

In manufacturing, lasers provide several unique advantages in respect of:

   (i) Superior product quality
  (ii) High productivity with reduced cost
 (iii) Greater material utilising
 (iv) Eliminating finishing operations.

The new modern industrial units would increasingly utilise laser technology on account of the significant and perceptible advantages it promises.

Revolution in Communication

(1) Internet

The Internet, a new computer-based medium, has brought about amazing revolution in regard to communications. Internet, which is mostly devoted to communication, is a collection of tens of thousands of networks of computers, spread across several countries of the globe. A menu based computer programme called 'world wide web' provides hyper text and hyper media links, to other information sources throughout the Internet. Users are now in a position to focus attention on fast-breaking news items, timely features constantly being updated, and referenced with previous articles and websites. One can get information on the latest developments in science and technology occurring in any part of the world, and utilise them effectively for research programmes. With the availability of Internet, dissemination of information takes place amongst several users at a very fast pace.

(2) E-Commerce

It has been estimated that approximately 10 million to 13 million e-mail messages are being transmitted in a minute. The E-Commerce business is projected to reach $ 7 to $ 8 trillion by the year 2004-2005.

(3) Views of Scientists/Technologists

Scientists/technologists associated with internationally renowned 'Bell Laboratories' (which has 11 Nobel Prize Winners to its credit) have made certain astounding predictions for the next decade.

(i) A mega network of several networks, would enfold the earth and produce a 'communication skin' with universal connectivity.

(ii) As a result of inter-connected devices, the volume of infra chatter among communicating machines, will surpass even communications taking place among human beings on planet Earth.

(iii) Even though business organisations of the 21st century operate in a world of uncertainty replete with several complexities, industrial consumer oriented services would be made available and firms can provide consistently high quality service after carefully selecting the customers.

(iv) Existing Internet would be transformed in due course to a broad-band HI-IQ NET with natural interfaces with active websites, and with software agents to extract the required information/data via. text, voice, images as well as video.

Fibre Optics

Several developing countries have started using fibre-optic – the new technology of transmitting information with beams of light to solve the multifarious problems in telecommunications industry and medicine.

As a result, telephone conversations, television broadcasts, computer information etc., or any other message could be translated into light wave, and sent through glass wires, instead of the conventional technique of translating them into electrical impulses and sending them through copper wires.

It is indeed a marvelous engineering breakthrough, that makes it possible to carry 100,000 phone calls or more than 100 broadcast signals on a flexible glass strand-the size of a human hair!

In the course of a decade, a single fibre optic cable may bring telephone, two-way television, video text information and electronic newspaper into home communication centres.

Improving Quality of Life

By 2020 in several countries technological changes would help the people to improve the quality of their lives on account of:

  (i) Changes in factory size: Options for smaller plants will be great
 (ii) Quiet/less polluting vehicles
(iii) Safer road surfaces
 (iv) Sophisticated insulated homes
  (v) Cheaper air travel
 (vi) Greater use of mobile-personal telephones
(vii) Revolutionary changes in transmission
(viii) Multiplicity of special interest TV channels

Employment Opportunities

There may be several new avenues for the performance of specific tasks and payment would be mostly on: piece work basis.

Winning the Battle in the Market Place

The Electronics revolution may facilitate infinite availability of information. However, the danger of poor quality information may also pose threats to correct decision making.

Creating wealth out of knowledge is the greatest challenge facing the developing countries of the world. The process of wealth creation is greatly interconnected to the mastery of manufacturing technology. Therefore, any developing country keen on accelerating the process of economic growth should be able to tackle effectively the multifarious challenges posed by manufacturing technologies.

While focusing on manufacturing technology, it becomes clearly evident that greater efforts have, therefore, to be applied for creating a really powerful workforce comprising 'knowledge workers', who are capable of converting knowledge into wealth. This would ensure high performance standards, besides enhancing the quality of different products turned out at a blazing speed. This would certainly ensure that the modern firm stays ahead of its competitors and wins the battle easily in the competitive market area.

Benefits arising out of science and technology, are measured not in terms of new knowledge, but by the capability of a country's science and technology to create and turnout new products and new wealth which would ultimately enrich the lives of the people.

During the short period, any industry in the realms of challenges faces:

   (i) Quality
   (ii) Cost
   (iii) Productivity
   (iv) Innovation and
   (v) Response time as regards utilisation of a specific technology.

Technological Competence

While in the long run, challenges will mostly centre on:
(i) The appropriate scale of operations
(ii) Technological upgradation
(iii) Induction of latest product designs
(iv) Benefits arising out of 'strategic partnerships' with countries well-versed in specific technologies.

Need for Diffusion

Different kinds of systems are in operation in different advanced countries of the west for the diffusion of technology. Different methodologies are adopted for such diffusion in Germany, UK and France. Different types of arrangements confer benefits on large scale, medium scale and small scale industries in these countries.

The German Vocational Education System has installed a 'massive system of technology information diffusion', which encompasses firms, industry-associations, experts, government departments and so on.

Arrangements are available for undergoing formal instruction by students/entrepreneurs in specific fields and the government takes responsibility for curriculum standards, examinations, certification, etc.

Innovations

Akio Morita of Sony Corporation has aptly stated that innovation is likely to happen only with 'visionary leadership'. He has remarked, "The innovation process does not emanate by bubbling up from R and D laboratory. It begins with a mandate which must be set at the highest level of the Corporation." The induction of Sony's famous 'Walkman' could be cited as an wonderful example of an 'innovation' inspired by the top man.

Besides, it could be stated that consumers and market forces could be attributed to the emergence of several innovations in several countries where they are the: 'drivers of innovation'.

The renowned Unilver Company strongly believes that as such 'innovation' is closely interconnected with business strategy, identification of consumer needs and so on. Identifying consumer needs, therefore, becomes 'integral' to the process of innovation, and ultimately it has much impact on the manufacturing and product technologies.

The 21st century business scenario demands high standards of product excellence combined with low cost and quick delivery schedules. In order to achieve these objectives, firms opt for the latest advances in chemical engineering, mathematics, physics, mechatronics and so on, and as a result new market opportunities could be immediately explored with induction of improvements in quality in a brief span of time. Consequently, an aspiring firm with sophisticated technology could forge ahead of all others, by presenting a product of unique quality and high performance standards, which in all probability would exceed the expectations of the consumers.

Need for a 'New Economic Paradigm'

The healthy interaction between the tremendous advances made in information technology and economics has been the subject of intense debate in recent years.

We are confronted with the peculiar problem of rapid strides being made in computer processing capabilities on the one hand, and the shrinkage in costs in several avenues which would provide a mixed bag of advantages and tricky problems. While several people are pessimistic on the resultant wipe out of jobs or decline in job opportunities on account of computerisation, quite a good number of them have expressed optimism over the possibilities of enormous growth in productivity over the years with only mild inflationary impact.

The forces of globalisation and breath taking advances taking place in technology are the most important twin factors, which have attracted the attention of economists, technologists and scientists alike, and they are now convinced that the old rules of economics are no longer applicable in a turbulent world.

The following viewpoints have gained increasing acceptance:

(i) The forces of information technology (computers, semi-conductors, software and telecommunication) unlike previous technologies would only be instrumental in destroying jobs/job opportunities on a vast scale.

(ii) Mobility of technology and capital, characteristic of the 21st century milieu, would induce the firms in rich, developed countries to opt for collaborations, and fixing up locations depending upon the 'cheapest area' as per their assessment.

(iii) Even though free trade, and advances in technology have liberated several economies from 'capacity constraints', the rigorous monetary policies pursued by the central banks of developing countries are inhibiting rapid industrial growth.

Therefore, the changing economic scenario in several developing countries call for a 'new economic paradigm' that would accelerate growth, taking into account the factors mentioned above.

Technological Abilities

It is an acknowledged fact that we are constantly engaged in multiplying our technological abilities at a faster pace. In the forthcoming decade, new businesses would mostly rely upon 'emerging new technologies', and traditional businesses will go out of recognition. As a result, the life style or work culture would be considerably modified.

In the ensuing years, hardware and software will acquire much sophistication and the location of the office (whether at a remote place or at home) will have little significance. Ultimately, management structures would be driven by information technology only, regardless of the location of the office premises.

Bill Clinton, the erstwhile American President, has aptly stated that an individual has to be 100 times 'smarter' than he was 10 years ago in order to achieve success and recognition at present.

Today, an entirely new battalion of professional IT managers is springing up, and dominating the business scenario.

Concept of Lean Production

The term 'lean production' introduced by the Massachusetts Institute of Technology has found universal acceptance. It insists on producing 'small numbers of customised products' at a phenomenal speed. The approach emphasises the importance of team work, flexibility of both equipment and people, and a commitment to total quality-all aimed at providing a rapid and efficient response to meeting customers needs. Besides, the firm would also be striving to 'minimise' the resources required to achieve the objectives: less equipment, less people, less space and less time and effort to develop 'new products'.

It will provide an opportunity to get closer to the ideal of manufacturing only against customers orders within an acceptable time frame and thereby eliminating the need to hold stocks (viz., inventory). The concept envisages, in addition to providing a rapid and efficient response to meeting customers needs, also including the design and development of new products, tailored and presented to cater for the specific consumer requirements.

'Lean production' meticulously follows a typical 'world class approach' and is gaining much popularity throughout the world.

An Ideal Manufacturing Continuum

In the current milieu, for the modern firm 'areas of opportunity' lie ahead of as well as on completion of manufacturing the continuum as presented in figure 3.1.

Technological Competence

Figure 3.1: An Ideal Manufacturing Continuum

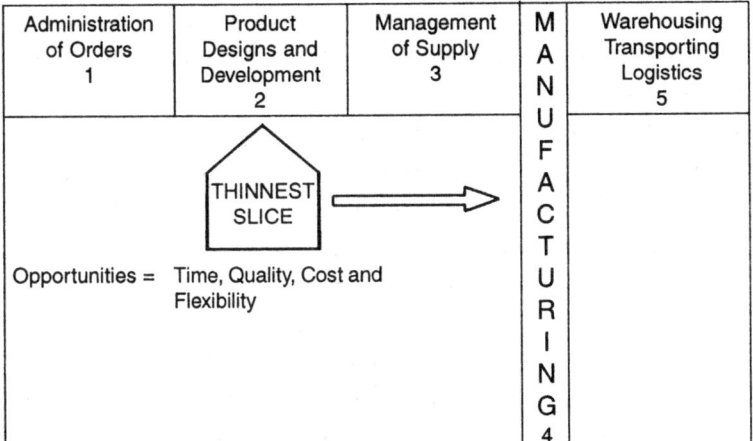

There may be possibilities of waste areas in terms of time spent and other factors in second, third and fifth tiers.

When we consider manufacturing (as per the linear representation in the figure) as commencing from the request by the party to shipment, it would be obvious that *actual manufacturing* (segment 4) would constitute the 'thinnest slice' of the overall process.

Smart managers have managed to prune much labour and waste out of the process. Semi-automated mills and robotic plants would also be utilised.

The current focus pertaining to 'lean manufacturing' is almost an obsession to drive away waste and complexity. Every bit of information from precise billing to correct product specification would be transmitted electronically.

Present day technology has achieved compression into a 'single input and output flow-circuit-diagram' (without linear system) capable of presenting: 'innovation in design and clean, fast replication!' Today's technology machine has gone to great heights and provides unthinkably high levels of performance at unimaginably low costs.

In recent years, a fundamental shift has occurred in the economic behaviour of information technologies. Previously, tasks were automated piece meal, and each automation project was assessed on the basis of individual merits. The benefits derived from the above approach were easily measured.

With the growing importance of communications technology, benefits may flow from automated applications by adopting systems-wide automation.

At the factory level computer integrated manufacturing systems could produce highly substantial economic gains, which would ultimately squeeze out excess inventory.

At the office level computer integrated systems could considerably reduce unproductive time spent by white collar workers.

Gaining Strategic Advantage: Prof. Humble's Viewpoints

Professor Humble has expressed the view that a technologically driven corporation could gain 'strategic advantage' only by inducting certain modifications as outlined below:

(i) Since cost of IT systems escalate, and risk factors are substantial, technical experts and general managers will have to cooperate with greater involvement in order to achieve professional excellence.

(ii) As such IT systems do not work well in the (old) "Classic Command and Pyramid" model. Only in flattened organisational structures IT systems function extremely well.

(iii) The normal reluctance of middle managers to share information would no longer be applicable. There is a need for greater rapport among different groups – 'since knowledge is power'.

(iv) An organisation having sharply defined boundaries is unrealistic.

As regards just in time delivery system, the company's suppliers are closely knit into an IT-web.

(v) Modern information systems provide enormous opportunities for sharing knowledge and extending learning process. From any part of the world one can have access to past experience of others, and thereby could learn useful lessons.

Expanding Intellectual Horizons

The renowned management wizard Peter Drucker has emphasised repeatedly that the modern executive has to learn continuously in order to expand his intellectual horizons in a highly dynamic business world.

All modern executives should possess precise and useful information covering aspects such as:

(i) What information they need
(ii) From whom
(iii) When it may be required
(iv) What information they can disseminate
(v) The time span to be spent and so on.

Therefore, modern executives should be constantly on the alert to 'learn' not only 'on the job' but also from management centres, industrial seminars etc., the latest changes taking place in the industrial scenario.

Michael Porter's Warning

The modern executive/manager could formulate an useful success recipe for his firm, only if he is in a position to gauge the newly available technologies being experimented by rival firms.

It is quite possible that the innovation on which a particular firm is relying upon might prove inferior to the one being taken up for launching by the competing firms. In this regard, Michael Porter has aptly warned an aspiring firm against any kind of 'inertia' in switching over to 'superior technologies'.

Needs of the 21ˢᵗ Century

(i) The IT department of the modern firm should come to the forefront of action, instead of functioning methodically, remaining in the basement.

The CEO as well as senior managers should allow 'Internet' and 'E-Commerce' to become an integral part of business strategy. Besides, they should try to explore different avenues by which IT could 'reinvent business' and confer benefits by way of identifying new opportunities thrown open to the firm. In addition, they should endeavour to chart out a 'long term growth path'.

(ii) 'Knowledge workers' should constantly update their knowledge and prepare themselves to handle the job of middle managers by taking keen interest in the firm's activities.

(iii) Though several achievements are possible through IT, the 'personal touch' of face to face interaction, through human contact, should not be forgotten. The all round quality of life could receive a boost only through healthy human interactions.

(iv) On account of 'globalisation', electronic media would enhance and accelerate the 'flow of technology' around the world. There would also be increasing availability of innumerable technical products manufactured in different parts of the globe.

In this regard, a typical modern firm should endeavour to carry out 'strategic R and D activities', in a place very close to the firm's geographic headquarters, so as to maximise benefits.

The laboratories should not be large enough limiting its employees within 400 to 700 nos, should function near the manufacturing plants and marketing departments of the modern firm in order to accentuate R and D efforts and reap greater benefits.

(v) As in Japan to effectively monitor the latest fashion trends, consumption pattern, consumers' tastes and so on, units could be set up to provide essential 'feedback information' to the modern firm.

(vi) Tapping R and D talent: In small countries like Sweden, multi-national companies can not hope to staff their R and D laboratories only with Swedish experts/graduates. In order to enhance the effectiveness of their R and D efforts, approximately 40% of highly qualified experts are therefore appointed from different parts of the world.

Besides, in order to attract highly talented specialists from across the globe to carry out R and D efforts successfully, special incentives and salary packages are announced by leading firms.

For the modern firm keenly aspiring to achieve excellence and outstanding success in the international arena, there arises the supreme need to 'attract the best talent' from all over the globe after meticulously scanning the environment.

Latest Developments

(i) The potential for innovation in manufacturing has acquired considerable significance in recent years. Advanced manufacturing technologies have thrown open several options to sourcing manufactured components and products from different countries of the world.

(ii) CEOs and technologists all over the world have noticed that the potential and application of 'Robotics' and complete 'Computer Aided Design' are growing approximately at a rate of 25% to 35% per year, and as a result there would be considerable boost in the sale of products emanating from the use of advanced manufacturing technologies.

(iii) Entrepreneurs are realising that their main concern would be not in making sizeable investments in new technologies, but it would centre on the problem of selecting the most appropriate technologies suitable for specific applications, and which are best suited for their business.

(iv) The shorter life cycle of several products, as well as manufacturing processes are much instrumental in price erosion varying from 20 to 25% per year, and as a result difficulties are keenly felt in building a sound and prosperous business, amidst increased world competition. In recent years, enterprises manufacturing personal computers, cellular phones, etc., all over the world are mostly affected by such phenomena.

(v) The modern technologically intensive firm should constantly be on the look out for recent developmental activities being undertaken in the realm of technological innovation in different parts of the globe. They should be relentlessly curious in regard to 'external sources of technology' and should endeavour to undertake rapid scanning, before making necessary efforts to induct relevant expertise originating in foreign laboratories and technical institutions. The aspiring enterprise should, therefore, rapidly assimilate relevant flow of 'technology expertise' across the borders and make every possible effort to replace out-of-date technological devices as fast as possible.

In accordance with global requirements, the modern firm should 'energize the innovation process' by implementing programmes based on 'technical market intelligence' gathered from different locations/countries of the world.

Latest technical know-how should be constantly injected into the system on the basis of its compatibility, irrespective of the country of its origin. R and D efforts and technological innovation process would acquire greater viability and dynamism, if up-to-date market intelligence could be made available in a vigorous manner.

(vi) Besides, the modern companies functioning in technologically rich environments should explore worthwhile alliances with leading firms in advanced countries such as: USA, Germany, Japan etc. This would enable them to import necessary R and D talent from abroad, in addition to facilitating enhancement of 'in house capabilities'. Combining with international expertise, such firms would be in a position to acquire the following twin advantages:

(a) They can maintain low cost of production.

(b) They can considerably upgrade the quality of the products being manufactured by them.

(vii) When we embark upon the 21st century business milieu, we are dramatically confronting different industry structures and configuration patterns, which are strange. Innovations are the outcome of the so called technologically convergent innovative products, emerging on account of the 'multi-dimensional interconnection of ideas'.

The emergence of the new 'media car' having an intelligent road navigator, an interactive entertainment centre, vide voice data communicator, a home from home, a playground etc., could be classified as 'hyper innovation'.

Several leading firms such as: Sony, Motorola, Nokia and Disney collaborate with giant firms like: Cadillac, Toyota, Ford and so on, and have succeeded in developing a new breed of 'infotainment-vehicles'. It is on account of integration of diverse technologies that the concept of growing a 'multi-dimensional enterprise is gathering momentum at present.

'Super food firms' are being organised to provide yoghurt with anti-depressents, and vegetables having antibiotic properties.

Cosmetics have been integrated with *pharmaceuticals* so as to offer shampoo that could enhance hair growth, besides highly effective *anti-ageing creams* and so on.

(viii) It is quite possible that complex innovations originate from lesser innovations. The foremost theme on innovation lays accent on 'many potential interconnection of ideas'. It is an acknowledged scientific fact that while we observe what emerges from a complex system of ideas, we find that the output or whole activity is greater than the sum of individual agent: $1+1+1+1 = 6$?. (It is "synergy"-the secret being 4 agents in a network have 6 possible interconnections). For every new scientific discovery, and resultant technology coupled to a network, the number of 'innovation possibilities' dramatically increases.

Dr Mitchell Waldrop has given a sophisticated scientific perspective to a novelty/innovation in his publication *'Complexity'* as given hereunder:

"It is essentially meaningless to talk about a complex-adaptive system being in 'equilibrium'. The system can never get there!

"It is always unfolding, always in transition. In fact the system 'never' does reach 'equilibrium'. It is not stable. It is dead ! And by the same token, there is no point in imagining that the agents in the system can ever 'optimise' their fitness, or their utility or whatever. The system space of possibilities is too vast: They have no practical ways of finding the 'optimum'. The most they can do is to change and improve themselves relative to what the other agents are doing. In short, complex-adaptive systems are characterised by 'perpetual novelty".

From the above, it is understood that frigid, isolated systems can neither anticipate nor perpetuate 'innovation'.

The logic is this: In the past, the reflex has been to stamp out 'novelty' but suppressing 'novelty' only destroys the very mechanisms that keep the system alive and well viz.:

(a) Business
(b) Technology
(c) Markets
(d) Customers

When a system is characterised by little or no change, it would mean the system is in 'equilibrium', and has reached 'near death' state.

Though for a good manager equilibrium may signify high level of order and certainty, and the objective to be achieved, it can not last long. The firm if left alone would have totally gone out of touch with 'live competitive environment'. It has to anticipate and react to 'new market permutations', and should not remain blind to new developments constantly taking place.

For example in previous decades the pursuit of equilibrium by firms such as: IBM, Xerox, Kodak, Chrysler etc., have landed them in a static condition, with no further improvements.

Besides, the quest for reaching 'optimisation' is ingrained in the mindset of several managers. Though it appears to be quite logical, it proves highly detrimental to harvesting a good number of 'innovations' by the modern firm.

In the modern world continuous improvement of prevailing products alone would no longer hold any fascination. The firm could remain successful only if aggressive tactics are adopted for turning out products having perpetual novelty. The founder of Sony had aptly stated, "If we want to sustain value in the market, we have to become our own 'best competitor'."

In this context, the imperative need arises for constantly 'reinventing the firm' within 4 to 5 years of duration, and this would mean-

 (i) Executing unaccustomed operational activities.
 (ii) Inducting new/talented people
 (iii) Inducting diverse technology
 (iv) Introducing fresh functions considered essential after proper reappraisal.

Though variations and turbulence through experimentation would impart the requisite dynamism and enrichment to the firm, too many changes would also prove harmful.

Therefore, the modern enterprise should endeavour to function in a highly competitive spirit imaginatively adopting perpetual novelty so as to succeed in the business-world of constantly shifting patterns.

Benchmarking

'Benchmarking' is concerned with intensive search for 'best practices' in vogue, irrespective of the source and making every possible endeavour to attain superior performance. Such an approach essentially involves continuously evaluating a firm's products, services and practices both against leaders in the industry as well as other competitors.

Certain Important Attributes

 (i) Here firms are presented with targets and not solutions, which automatically increase the aspect of ownership.
 (ii) It helps to create a uniform response from all departments of a business firm; since goals set up are common, achieving improvement becomes a shared task.
 (iii) Benchmarking helps to reinforce the executive role all around, and facilitates sustained performance improvements.

(iv) New improvement horizons are opened up which enable the firm to step up performance. As a result, an opportunity to 'leap frog competitors' would be created in regard to several dimensions of performance.

Implementing Benchmarking

(i) Firms should fix up suitable targets which are high enough and rigorously pursue them. Targets should be on the basis of authoritative knowledge, and not on intuition or guess work.

(ii) An essential prerequisite would be to instil among staffers the sense of responsibility and accountability necessary to achieve superior performance.

(iii) There should be 'essential reorientation in approach which' would lay accent on managers focusing greater attention on external changes/improvements taking place in the industry, rather than confining much attention to internal issues.

(iv) In addition to assessing themselves against externally derived standards, firms should also take care of:
- Internal benchmarking by studying other parts of the firm
- Direct competitors in the industry
- Leading companies which belong to the same industry, even though they are not direct competitors.
- Latent competitors
- Relevant useful ideas gathered from firms outside the industry.
- A group of talented people could be appointed as consultants or hired until the successful completion of a specific project.

In the present day turbulent and highly competitive environment, a modern aspiring firm has to guard itself against the 'capsizing effect'. Planning to recover from being out-performed and strategically out-manoeuvred should never be relied upon, in the present milieu, even though in previous decades such recoveries were possible.

The modern firms should, therefore, be pro-actively alert in today's highly competitive market environment. Indeed, lack of sufficient insight, and corporate awareness about competition creates vulnerability, and ultimately the 'capsizing effect' (sinking) overtakes. The modern firm, should never allow the other firms to competitively outmanoeuvre its position.

4

Customer Focus and Global Marketing Strategy

"Individual customers are today better educated and better informed than their parents and grand parents. Added to this customers have personal 'Dreams' and 'Aspirations'-which go far beyond the modest expectations of previous generations."
—Jane Smith, Consultant

In the 21st century business environment, there arises the imperative need to challenge the deep seated *business paradigms* already existing and for adopting new ones found more appropriate. 'Strategies' and 'systems' were hitherto built around the products being turned out by a manufacturing concern.

In this age of accelerating change, increasing competition, and instant communication, business people all over the globe have realised the need to build strategic enterprises by designing their businesses around customer types by assigning greater importance to the relationship formula. The modern 21st century firm is inclined to decide the entire business from top to bottom around a clearly defined 'type of customer!'

The greatest challenge confronting the modern firm is finding a suitable road map that would transform it to a customer centric strategic enterprise. Nowadays, any firm aspiring to achieve spectacular success in the international arena should no doubt make every possible endeavour to modify its 'strategies and systems', which were previously built around products. Besides, it is also absolutely imperative to abandon all 'out dated thinking' connected with marketing programmes.

Though the organisation's overall performance is of paramount importance for its success, the central executive officer (CEO), and other executives should also focus critical importance on how deftly the organisation is treating its customers while effecting sales.

Definition of Customers

Only on the basis of customer status, most of the firms classify 'customer segments.' One useful way of classification could be:

(i) Current Customers
(ii) Former Customers
(iii) Competitors' Customers
(iv) Those who use substitute products or services.

Every firm considers it most appropriate to focus attention at the outset on current customers, since keeping an existing customer is always found advantageous, and cheaper than identifying a new one. However, in the interest of expanding the market, the firm has the prime necessity to understand the needs of other 3 categories of potential customers as well. There may be subtle differences in regard to the needs of these potential customers, and this problem could be solved by effecting minor modifications in the products/services being offered by the firm.

It is an acknowledged fact that customers constitute the life blood of any business venture. Regardless of the nature of business only customers generate the income required to carry on business. In the present milieu, any complacent firm which merely satisfies or meets the expectations of its consumers in all probability would go out of business quite soon.

To make a lasting impression on the consumer, the firm has necessarily to offer a product or service which would exceed the customers' expectations. For the modern firm, the strategic requirement would be delivering service that is *fast, imaginative, excellent and customized*.

Only by implementing a strategy of highly effective exceptional quality service, the firm can hope to remain in the forefront for ever. Ultimately the companies which survive and thrive are the ones which are 'willing to listen, to learn and quickly act.'

'Delighting the customer', though appears simple on paper, is the greatest challenge faced by any modern firm. The success and prosperity of most businesses depend on the capability of the firm to maintain the loyalty of several customer groups. Besides making every possible effort to satisfy the requirements of external and internal customers, it is utmost vital for a thriving concern to create close alliances with suppliers of resources and raw materials in a sensible fashion.

The modern organisation could out-manoeuvre competition only by retaining existing customers and by constantly winning new ones. This would involve managing every part of the supply chain to optimum effect.

Customer Driven Operational Strategy

To achieve phenomenal success in the field of *consumer focus*, the modern firm should incorporate certain principles in the formulation of suitable strategies as outlined below:
1. Commencing operations with a 'specific customer type'.
2. Endeavouring to provide 'unique value' before making aggressive efforts to compete with rival firms.
3. Providing unique value in products with distinctly fine and extra ordinary features could be effected by encouraging team work wherein each employee gets the opportunity to experience every job in the production sequence.

Each individual employee understands better, how he fits into the system, and how his performance would affect the quality of the products. Besides, all employees are allowed to mingle as 'one team' having pride, enthusiasm, and greater commitment. They directly establish contact with people, and ascertain from them their likes and dislikes in regard to the specific features of a product, and how it could be improved so as to provide utmost satisfaction. By delivering unique value to designated 'customer type', the modern firm not only improves customer satisfaction, but also increases the perception of quality associated with the product or service provided. Thereby, the organisation's creativity will not be enslaved by 'product centric thinking'! In a typical product centric organisation, the manufacturing department, may use one type of information system, while the marketing department may utilise altogether another set of data. In such companies team work is sadly lacking, and mostly people, strategies as well as systems... do not work together. Therefore for a product centric company adopting a fragmented approach, delivering 'products possessing unique value' would be almost impossible.

Besides, it is utmost necessary to envision an *'ideal system model'* having specific objectives and goals for toning up overall performance. The company should formulate a detailed blue print for facilitating 'ideal outcome'. Instead of envisaging an all encompassing model, if the firm opts for the incremental method of making small step by step improvements in the existing system, it will have very little positive impact on the overall fortunes of the company. The incremental method proves ineffective in the long run, since the top executive has no courage to make radical changes to implement an 'ideal system'.

In each country, the socio-economic fabric is constantly changing at a rapid pace. In this age of accelerating change, the modern manager has no alternative except discarding the incremental method of making improvements in his firm. He should constantly look forward to a more desirable future, and therefore should envision an 'ideal system model' which would serve as a *light house!* He would be steering in the right direction, inspite of the fact that the task of achieving substantial progress may take a longer time than the time taken for short-term improvements.

Dr. Belbin's Team Approach

Dr. Meredith Belbin, a renowned psychologist had propounded the 'team role' theory based on his research findings that "imperfect people can make perfect teams" through balancing weaknesses with strengths in the team make up.

He had arrived at the conclusion: "Diversity is in our genes. It is how to capture this diversity in groups and use it to best advantage".

According to him, *complimentary contributions* produced "better results" than *competing contributions* and his key finding was that 'certain strengths' in individuals were often associated with 'certain allowable weaknesses'. Such weaknesses were not detrimental to effective team work because they could be underpinned (strengthened) by the strengths of others. He further stated that "sharing this knowledge helped to build effectiveness" along with the realisation that "imperfect people can make perfect teams".

He enumerated the following main roles which would contribute to form an ideal management team:

(i) **Company Worker** – Dutiful with organizing ability, hard work

(ii) **Chairman** – Self-confident with capability to treat all potential contributors on their merits with objectives.

(iii) **Shaper** – Outgoing and dynamic; readiness to challenge inertia, ineffectiveness and complacency.

(iv) **Plant** (Staff) – Individualistic, serious with genius and imagination; weakness → disregard for practical details and protocol.

(v) **Resource Investigator** – Highly extroverted and enthusiastic with capacity for contacting people and exploring new avenues. Capable of responding to challenges.

(vi) **Monitor-Evaluator** – Socially-oriented, mild and sensitive with ability to respond to people and situations; promotes team spirit.

(vii) **Completer-Finisher** – Highly painstaking, conscientious and aims at *perfectionism*. Though all these categories of people may have certain inherent 'allowable weaknesses', by and large their efforts would generate substantial benefits to the firm.

An additional advantage emanating relates to "integration of self-perception and perception by outside observers". Really a congregation of enthusiastic individuals could produce better results by tackling complex problems, which are periodically being confronted by the firm.

Dr. Richard J. Schonberger's Approach

Dr. Richard J. Schonberger, called the evangelist of 'customer driven performance', has propounded the concept: "building a chain of customers", which he claims as the ultimate theory to reach *world class excellence*. It is applicable in manufacturing as well as service industries, and such excellence could be achieved by regarding each function in business as a customer of the one serving it. In his theories the many links between and within the different functions of a business-design, manufacturing, accounting and marketing – form a continuous chain of 'customers' leading directly to those who buy a final product or service.

Schonberger has distilled his foremost ideas into a mission statement: *"World Class Excellence is Continual Improvement"* in serving the customer's four basic wants:

1. Ever better quality

2. Ever lower costs
3. Ever increasing flexibility
4. Ever quicker response

Tom Peters, one of the distinguished management experts, has greatly admired the concept put forward by Schonberger by saying that "it is a bold and meticulously detailed blue print for redesigning operations to destroy functional myopia to live as a whole to serve the customer". Schonberger's customer focused principles have been gleaned from some of the world's renowned manufacturing companies.

The concept of 'cellular manufacturing' is the foremost tool being employed by Schonberger to the 'internal customer chain.' Here clusters of people and operations are suitably arranged according to work flow rather than by departmental requirements.

Adopting Prof. Schonberger's principle, the entire Toyota family of companies suitably modified to move all its machines into 'cells', and built all its new plants according to the way the product flowed rather than by departments.

Invigorate with Matchless Customer Service

In present day environment, customers are mostly well-educated, and have wide travel experience. As a result they maintain an improved frame of reference, while purchasing different products for specific services.

(i) Invariably, manufacturing firms, until now, were making several exaggerated claims relating to the special features of their products. Such claims raised the expectations of the consumers, and later when the firms failed to deliver the required goods as promised, the consumers were much disappointed. Such an approach had produced adverse effects on the healthy growth of the firm in the long run.

(ii) In the present business environment, a firm keen on providing remarkably good services should adhere meticulously to the high standards expected by the consumers, and therefore, the strategy adopted should provide eloquent testimony to its impeccable service norms.

(iii) In the 21st century milieu, when greater accent is laid on 'time management', customers, in addition to evincing keen interest in getting better quality products and better services, insist on delivering the correct orders on or before the promised date. Therefore, the organisational culture must be well attuned to the 'customer defined standards'.

The Japanese Ideology

During interaction the Japanese are reported to be quite gracious and hospitable; but they are quite hard and aggressive, when they handle business transactions. In addition, every Japanese employee believes in achieving the goal: *'ichiban'* – which means number one-the biggest and the best. These are their ingrained habits, and they have achieved great success by inculcating such ideas.

(i) A modern firm keen on achieving rapid progress with a wider perspective should formulate an appropriate programme to investigate the essential requirements of the customers. By checking constantly on what the customers want, and by creating a system that would achieve the objective of fulfilling their requirements, a firm could accomplish much and remain always in the forefront.

(ii) McDonald's French Fries are renowned all over the world for excellent taste and consistent quality.

Besides devoting all the time and attention to business,

McDonalds had meticulously planned the best ways for growing potatoes, ensuring effective storage and by employing professional cooks to ensure smart cooking. Thus, McDonalds had virtually guaranteed that consistently excellent quality French Fries would be made available to the customers.

(iii) As regards 'Walmart', it has an exceptionally fine inventory system. It keeps on record all the fast moving items as well as items sold every day. Replacements are instantaneous and the orders are being executed by 'Walmart' immediately after receipt the next day.

Besides reducing its inventory costs tremendously, Walmart's systematically delivering consistently excellent products (garments/sweaters, etc.) at a short notice, had earned the firm a great name in its product line.

Carl Sewell, the renowned industrialist of Dallas – Texas, USA has aptly stated: *"The most important thing to a customer is: Did you do what you promised. Keeping your word is worth more than all the empathy, smiles, and chocolates on your pillow in the world."*

Customer Retention

Customer retention concept would become a reality, only when in an organisation employees having the right attitudes are given due importance. Such an organisation should ensure that it recruits, trains, manages and suitably rewards people, so that all the staffers enthusiastically carry out the things that are necessary to the customer.

M/s. Sears-Roebuck of USA formulated the following model, which brought about changes in the employee attitudes and thereby achieved greater customer satisfaction, and perceptible improvement in financial performance.

The following Figure 4.1 depicts how greater 'customer satisfaction' is being achieved.

Figure 4.1: Greater Customer Satisfaction

INCREASED EMPLOYEE SATISFACTION AND LOYALTY leading to
HIGHER QUALITY SERVICES AND PRODUCT EXCELLENCE leading to
INCREASED PROFITS AND GROWTH Leading to
INCREASED CUSTOMER SATISFACTION/ LOYALTY leading to

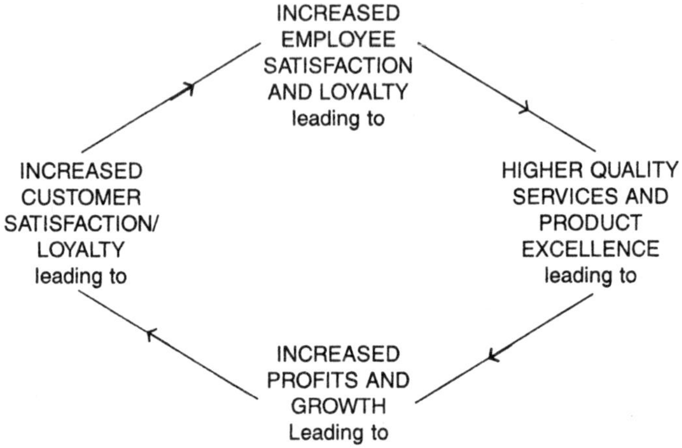

Kano's Model on Customer

According to Dr. Noriaki Kano of Science University, Tokyo, there are 3 different categories of customer perception: *(i)* Must be, *(ii)* More is better, *(iii)* Delighter.

Figure 4.2: Kano's Model

CUSTOMER SATISFACTION
DELIGHT
DELIGHTERS
NEUTRAL
MUST BE
DISSATISFACTION
ABSENT <--> FULFILLED
Presence of Characteristics

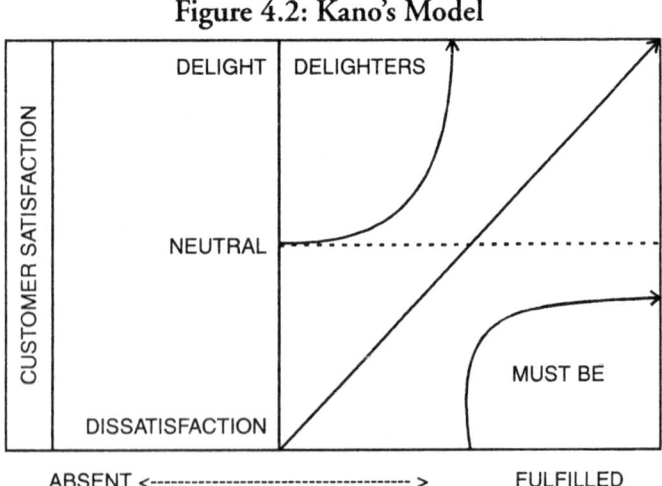

(i) **Must be:** Such characteristics are taken for granted like provision of hot water, clean linen etc., in a hotel room. They are normally expected and customers are annoyed due to their absence while the presence of these items would be treated neutral.

(ii) **More is better:** When a need is poorly met, customers feel disappointed. When queries result in instantaneous response, it may delight the customer! Something 'in between" (in this case) will evoke no response.

(iii) **Delighter:** Certain characteristics and features solve specific needs of customers even though they do not expect these to be available. As such, there is no negative effect, if it is not available. But when present, they have a positive effect. *For example:* separate study table/chair/lamp fittings, remote control for TV, provision of liquid soap, stereo cassette for music in a hotel room.

How the Model is helpful ?

1. It is Much Helpful to Employees to Set Priorities

Even at the hotel room, one can get rid of things that are disappointing. Guaranteed reservation' is a 'must'. Later visitors can turn attention to 'more is better' features and 'delighters'.

2. Avoid the "TRAP"-'No Complaints' Equals Customer Satisfaction

Companies tend to calculate customer satisfaction after finding out customer complaints. Absence of a negative will not create a positive.

A Company, may not drive away customers; but it may not be able to provide genuine delight to create more loyal customers.

Accent on Customer Delight

John Kay of London Business School has provided the following model (as per figure 4.3) in regard to core business elements which any global firm could advantageously adopt for achieving 'customer delight'.

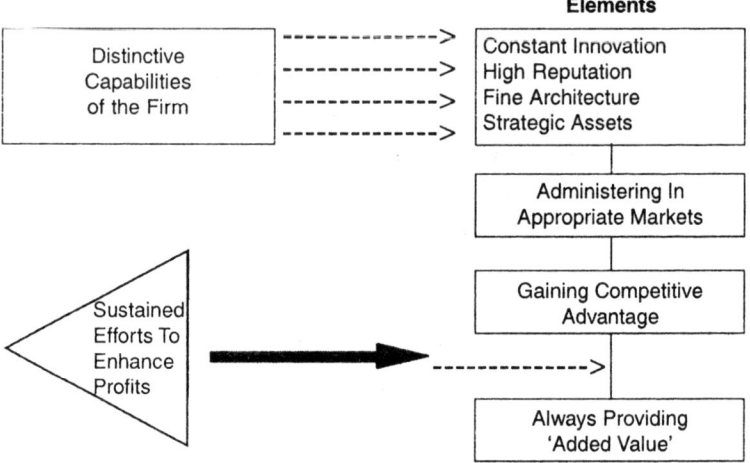

Figure 4.3: Model on Core Business Elements

Only by correct understanding of the genuine needs of the customers and by making concerted endeavours to provide these services, a global firm can hope to enjoy a prestigious position. It takes several years for the firm (for Marks and Spencers of the UK) to build up a combination of distinctive capabilities which help the firm to offer unique products that would captivate the consumers and constantly harness their curiosity.

Some of the 'customer defined standards' which will act as targets for the Global firm are the following:
1. Capturing the 'voice of the customer'
2. Reducing 'delivery time' to gain an edge over competitors
3. Taking up appropriate 'diversification' in a product line by offering a variety of products.
4. Provision of 'superb customer service'
5. Induction of 'new sophisticated technologies' to suit latest requirements.
6. Capability to offer 'tailor made products' *Example*: a car manufacturer should announce different varieties of

Customer Focus and Global Marketing Strategy

steerings, upholstery, seating, head lamps, and other requirements as per the style opted by the customer for a specific model sold at a specific price.

Focusing on the Outer Rings of the Total Product

Management expert Tom Peters explains how Harvard University's Prof. Tedd Levitt has provided a wonderful device (Figure 4.4) to aid systematic analysis of this prescription.

Figure 4.4: Focus on Outer Rings of the Total Product

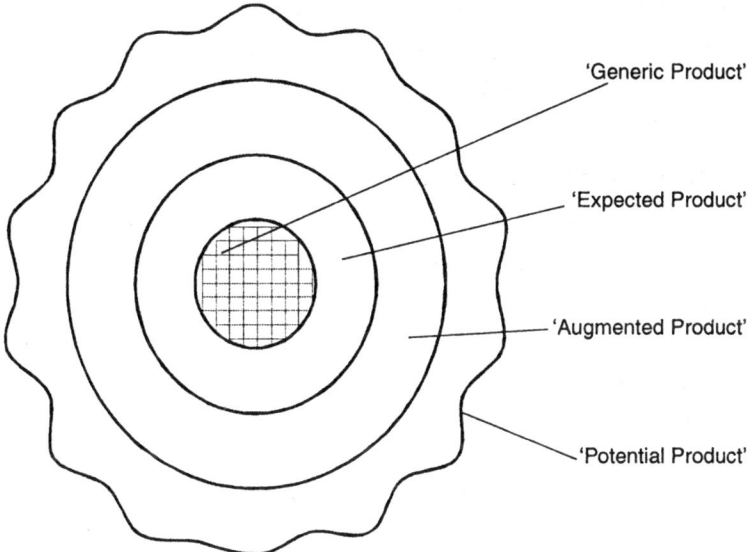

Example for Exhibit - 11:

Functioning of a Clothing Store

(i) *Generic Product:* At this level a clothing store regularly provides certain categories of top quality items.

(ii) *Expected Levels:* Extraordinary response to changing fashions and designs with constant updating.

(iii) *Augmented Level:* Overpaid sales people to ensure availability of more sizes, more colours and to the required tastes.

(iv) *Potential Level:* Fragrance in dressing room with flowers, in addition to facilities for returning clothes and providing alterations as per requirements and exceptionally fine acts of service.

Such a clothing store will certainly earn the reputation in the field of speciality clothing – where shopping is an experience by itself, a provider of a lifetime (needs) having great user friendly approach.

Every global organisation, in addition to paying much importance to the general level, should also devote much attention to repositioning the products and creating new markets by effectively tackling the 'two Outer Levitt Rings' also. This would bring about spectacular results in due course.

Global Marketing in 21st Century

With the advent of tremendously fast communication facilities, transportation network and quick financial flows, the world as such has started shrinking rapidly.

Products manufactured in one country find exceptionally enthusiastic acceptance in different parts of the world.

Montblanc pens, McDonalds hamburgers, German BMWs, Toyota cars, Boeing jets, Kodak cameras and Indian silk sarees are universally appreciated and purchased in the world market.

International trade which has steadily grown over the decade has attained unprecedented heights in recent years. In 1970, around 7,000 multinational corporations were making impressive strides. Now the number has more than tripled with 24,500 MNCs dominating in different categories of product lines and services all over the world.

In the USA and several other countries, Toyota, Sony, Nokia, Mercedez Benz, Panasonic, Nestle, Gillette and Colgate have now become 'household words'.

Day by day, competition at the global level is intensifying, and several leading firms are adopting aggressive tactics to remain in the forefront. Some firms endeavour to curtail foreign supporters by adopting protectionism. However, in the long run, this would bring about a rise in the cost of living and encourage inefficient domestic companies. Instead, they should come forward to effect continuous improvements in the products being turned out by them, so as to capture sizeable domestic markets as well as expanding foreign markets.

Firms which are highly hesitant in taking effective steps towards internationalising their business would have no opportunity to garner the wonderful benefits being presented to them by the glamourous markets growing within European countries, the Pacific Rim countries and so on.

By being 'over cautious' in approach, it is quite likely that such domestic companies may face tough competition emanating slowly from foreign firms being established in their home countries.

Firms, which opt for global operations, however, must guard themselves against certain 'unavoidable risks' such as:

 (i) Inflation/unemployment in the foreign countries may pose problems in addition to political/economic instability due to unstable governments and currencies.
 (ii) Strict regulations are announced by some countries for foreign firms and there are also regulations limiting the profits, which could be taken to their home countries.
 (iii) Several foreign governments in order to protect their own industries impose high tariffs and trade barriers.
 (iv) Since corruption is quite rampant throughout the globe, officials in several countries quite often show favouritism/ partisan attitude to the briber instead of the best bidder offering quality products.

In spite of the above mentioned shortcomings, several firms have to internationalise their operations in order to grow on healthy lines and reap greater rewards.

Prof. Gary Armstrong and Prof. Philip Kotler, the renowned marketing experts, have explained the terms 'global industry' and 'global firm' in a succinctly fine manner as follows:

Global Industry

"A global industry is one in which the competitive positions of firms in given local or national markets are affected by their global positions."

Global Firm

"A global firm is one that by operating in more than one country gains marketing, production, R and D and financial advantages that are not available to purely domestic competitors."

The 'global company' sees the world as one market! It minimises the importance of national boundaries and raises capital, obtains materials and components, and manufactures and markets its goods, wherever it can do the best job!

Here, Prof. Armstrong and Prof. Kotler cite two wonderful examples to substantiate the multifarious advantages accruing to a 'global firm' in the following lines:

   (i) Ford's 'world truck' sport, a 'cab' made in Europe, and a 'chassis' built in North America. It is assembled in Brazil and imported to the United States for sale.

   (ii) Otis Elevator gets its elevator door system from France, small geared parts from Spain, electronics from Germany, and special motor drives from Japan. It used United States only for 'systems integration'.

Thus the global firms gain advantages by planning, operating and coordinating their activities on a worldwide basis.

Regardless of its scale of operations, a global firm (whether large, medium or small) should consistently endeavour to assess and establish its place in the world market.

Several worthwhile factors such as the market position within the country and globally, the strategy being adopted by competitors, suitable locations for manufacturing specific products, components, etc., should be thoroughly studied, by the global firm in order to make correct decisions concerning its progress on healthy lines.

The global firm before taking up the decision to operate internationally should have thorough insight into the current international marketing environment.

A global firm is confronted with *'six major decisions'*, which have to be effectively tackled at the outset. The following figure 4.5 provides details on challenges being confronted and the decisions to be taken by the global firm:

Figure 4.5: Decisions to be taken by the Global Firm

1. Study the Present Global Market	2. Firm Decision to Go International	3. Selecting Suitable Markets
4. Methodology of Entrance to the Market	5. Working out a Global Marketing Programme	6. Constituting Global Marketing Network

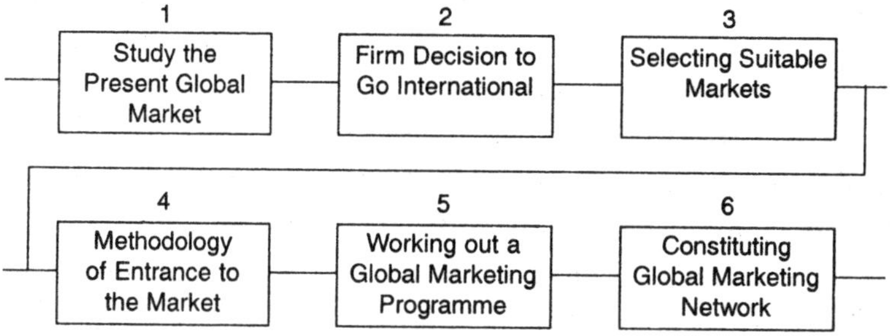

Study of the Global Environment

The global firm must be well-versed with the international trading system covering several aspects.

(i) *Tariff:* The tax structure adopted by a particular country (govt.) against certain imported products.

(ii) *Quota:* The exporter would be facing the quota system applicable which sets the limits on the amount of goods that an importing country will accept in certain product lines.

(iii) *Exchange Controls:* Global firms face exchange controls wherein the amount of foreign exchange that could be obtained and the exchange rates against other currencies are prescribed.

(iv) *Embargo:* Indeed it amounts to boycott which totally bans certain kinds of imports.

(v) *Non-Tariff Trade Barriers:* Restrictions imposed by a particular country and rigid standards being prescribed by certain countries for importing products.

Japanese insist that all foreign cosmetic companies must adhere to the stringent standards prescribed by them since the Japanese claim certain 'uniqueness' required in regard to their skin protection.

(vi) *GATT-Promoting World Trade:* The function of General Agreement on Tariffs and Trade (GATT), however, is instrumental in promoting world trade among nations and by reducing tariff and other international trade barriers. At present more than 140 nations come under the purview of 'GATT'

Since its inception in 1948, after several rounds of discussions, it has made a remarkable achievement in recent years by reducing the 'worldwide tariffs' from 45% to 5%.

The Uruguay Round of Discussions, which concluded in 1993, confer several benefits to promote long-term global trade. Besides reducing several merchandise tariff by 30%, it has helped much to boost global merchandise trade by 10% approximately. Moreover GATT has been extended to encompass trade in agriculture as well as in a wide range of 'services'. It has adopted a tough stand in regard to international protection of copy rights patents and trade marks as well as other intellectual property.

The Uruguay Round, besides reducing trade barriers, and prescribing international standards for trade transactions, has established the World Trade Organisation (WTO) which would effectively implement 'GATT-rules'. Its general functions are: (i) Hosting negotiations on general agreement on trade in 'services'; (ii) Activity as an 'umbrella organisation' supervising GATT-the general agreement on trade in services and 'agreements' governing intellectual property rights (iii) Mediating in global disputes. It also imposes trade sanctions that the previous GATT organisation never possessed.

Top decision makers associated with WTO meet once in two years to discuss the widespread ramifications of matters pertaining to WTO agreements. The last meeting of WTO was convened in 2001 at Doha in Quatar.

Each nation has its peculiarities and attaches considerable importance to certain unique features of products which they are keen on buying from foreign markets. In this regard, each country's economy has a distinct industrial structure which is different from others, and depending upon factors such as income levels, employment levels and so on, market opportunities could be well-ascertained.

Certain countries rich in natural resources like Saudi Arabia (oil), Chile (copper and tin), and Zaire (coffee, copper, cobalt) export mainly raw materials, and they invariably require large equipment trucks and tools to handle them.

In such countries, the rich and affluent constituting the upper strata of the society opt for luxury items in large quantities.

In the 'industrialising economies' like India, Egypt, Brazil, Philippines, manufacturing as such accounts for 15% to 25% of the total national income of their economies. They import substantial quantities of steel, heavy machinery and equipment, raw textile-materials and so on. A newly emerging rich class, and a steadily growing middle class in these countries create substantial demand for different imported items.

The 'highly industrialised economies', exporting large varieties of manufactured goods all over the world have large middle class segments which create enormous demand for multifarious sophisticated/utility products.

Since the political and legal environments of nations differ vastly, international marketers should ascertain at the outset the nature of political stability and legal structure of the country before entering into business and financial transactions.

There is also the imperative need to study the cultural environment of a particular nation before entering into any such transaction. There are vast differences in regard to fashions, fads, norms, and taboos of consumers in different parts of the world. Certain peculiarities which need greater attention are enumerated below:

Consumer Preferences:

(i)	France	The average French man uses more cosmetics, beauty aids than the average gentleman in any part of the world/drinks more wine.
(ii)	Italy	The Italian children are accustomed to large slices of chocolates.
(iii)	Germany	The Germans eat more branded high quality spaghetti.
(iv)	Tanzania	The Tanzanian women are afraid of giving 'eggs' to children.
(v)	USA	The average American drinks five times as many soft drinks as any European and consumes substantial quantities of poultry products.
(vi)	Brazil	The Brazilian mothers believe that babies could be fed by 'prepared food' and feel reluctant to buy processed foods.

Business Behaviour:

(i) The Japanese are averse to fast and tough bargaining approach; and during face to face conversations they are quite positive. Even in most Asian countries polite conversation is a must!

(ii) South Americans prefer business discussions in a cosy atmosphere in an informal manner.

(iii) American executives exchange visiting cards in a formal manner, whereas the Japanese study the contents: designation/firm, etc., and give due respect....

Going International

(i) The global company should safeguard itself from foreign firms which may try to offer better products at lower prices at the domestic market itself.

(ii) The global firm should make concerted endeavours to enlarge its customer base (in the domestic market and in the international market) to enjoy the advantages of 'economies of scale'

(iii) It should carefully weigh the multifarious risks involved in operating globally. It should be able to gauge the customer preferences and behaviour well in advance.

(iv) It should possess sufficient capability to adapt itself to the business culture of a foreign country and also deal effectively with foreign nationals.

Selecting Suitable Markets

Before entering the international arena, the global firm should formulate its 'marketing objectives and policies' in a well-planned manner.

It has to decide the types of countries which it would enter, and ascertain details of emerging new markets all over the world.

By and large, decision making is greatly dependent on the products required, geographical factors, income brackets to be served, total population, political factors, and socio-cultural factors in foreign countries.

The General Electric Company (GEC) has succeeded in selling approximately 12 to 15 million appliances/gadgets etc., in more than 150 markets spread all over the world. It studies meticulously and microscopically all details beforehand to assess the market growth potential in each country before embarking on its marketing programme.

Methodology of Entrance to the Market

While selling in a foreign country, the 3 foremost options open to the global firm are:
- *(i)* Exporting Products
    - *(a)* Indirectly or
    - *(b)* Directly or
    - *(c)* Through foreign based distributors
- *(ii)* Entering into joint ventures with reputed foreign firms
- *(iii)* Direct Investment:
    Entering a well-known foreign market by developing foreign based assembly/or manufacturing facilities.

Working Out a Global Marketing Programme

The global firms have two options:
- *(i)* Adopting standardised marketing mix which is cost effective.
- *(ii)* Adapted marketing mix where product preferences of local conditions are studied and given due regard.

Though standardisation helps a great deal in keeping down costs and prices and facilitates building up of an impressive global brand, it may not succeed much in the long run.

Marketing experts invariably make the ideal suggestion "Think Globally And Act Locally". They strongly defend the idea: A global firm can always *standardise core marketing elements* and localise others, and suitably blend them in order to satisfy different types of foreign consumers.

Constituting the Global Marketing Organisation

At the outset, a global enterprise should suitably re-organise its export department to ensure that it functions most effectively to handle international problems/exports.

When several international markets have to be served, it would be advantageous to install a separate international division with adequate staffers to handle key areas like planning, manufacturing, research finance and marketing. Competent consultants/specialists to tackle intricate problems could be inducted.

When global operating units are established in due course, the global units report directly to the Central Executive Officer.

Besides suitable training programmes will have to be organised for the benefit of executives, so as to help them think in 'global terms' and tackle worldwide operations most effectively.

Ultimately, the global firm may recruit its staffers/managers hailing from different countries depending upon their capabilities and track records.

A global firm may create separate departments to handle different products such as: toilet soaps, baby care products, hair care products, perfumes, talcum powder and so on in order to streamline the entire 'strategy' and offer 'innovative products' at a short duration.

Organisational Changes

"Keep Learning": "In a Technology Business, everybody has to acquire knowledge at a prodigious rate. At 'Microsoft' – we read, ask questions, explore, go to lectures, compare our notes and findings with each other, consult experts, day-dream, brain storm, formulate and test hypotheses, build 'models', and simulations, communicate what we are learning and practice new skills.

—**Bill Gates**

In the 20th century milieu, businesses were mostly characterised by manufacturing (oriented) units, catering mostly for the domestic economies. When we closely observe the present trends, organisations have become highly complex on account of their global operations.

Approximately fifteen years ago, the organisational design was almost a hierarchy resembling a 'pyramid'. It had a president, or CEO at the top, with a team of managers supporting him. To the managers, several supervisors were reporting, and at the bottom, only the employees were actually designing, producing and delivering products and services. Since most of the decisions were taken by the CEO or the vice president, the bureaucracy involved and its adoption of a time consuming, circuitous process caused much inconvenience to employees. They were constantly approaching the higher level executives in the hierarchy even for making minor decisions.

Later dramatic changes took place when the leaders of Granite Rock decided to turn the organisation chart upside down by formulating customer focused design, wherein the President functioned at the bottom! A few years later, this methodology of approach adopted by Granite Rock earned the prestigious 'Baldridge Quality Award' for the organisation.

It was increasingly being recognised that the organisational structure was built around people: who could enhance their work sensibly in order to provide the customers what they wanted, besides helping the organisation to earn substantial profits.

Prof. Mike Johnson has observed that there is a growing sense that people who have worked for thirty years in the same firm are too set in their ways, too parochial (narrow/isolated) in their culture. Life inside company will increasingly become the preserve of the few and the young!

He has further stated that people who form part of the top management should make honest assessments of their contributions as well as those of their colleagues by providing answers which truly explain the present nature of functioning... as outlined below:

(i) Whether the top management has formulated an 'integrated plan' which would:
- Cut down costs
- Reduce unnecessary staff
- Cut down inventories

(ii) Are they taking up seriously:
- Quality control and improvements
- Realistic assessments of the competitive edge of the product.
- Continuous breakthrough in the field of innovations.
- New marketing techniques
- Deepening existing relationships
- Making efforts to attract talents

After carefully analysing these factors, ultimately the challenge may turnout to be one of 're-designing the underlying management-processes' as a whole. For any organisation, in order to radically change and yield promising results, there arises the urgent need for fundamental transformation.

In this regard, Prof. Champy has aptly remarked that, "After the first surgery is over, executives must fundamentally transform orthodox management models', 'mindsets' and 'values' from those oriented towards 'command and control' to those that promote and lead an ennobled environment."

Besides, leading management experts and technologists have expressed the view that 're-engineering is not a project' having a programme with a beginning, middle, and an end. It is a 'process' that does not stop.

At present several organisations have opted for 'reorganisation' on healthy lines, since they are keen on surviving into the next millennium. However, just by concentrating on reduction in inventory by 50%, reduction in capital by 40%, and by reducing staff-strength, one can not expect dramatic performance improvements!

In this regard, every manager must use his creative potentialities to the maximum, and it is critically important that he concentrates on the 'creation' and pursuit of opportunities' being presented by the changing socio-economic scenario. In order to achieve strategic objectives, the manager must make use of better strategies and improved organisational capability.

'Improved organisational capability' would mean, in essence, improving the capability of the organisation to design, generate, support and deliver the results contemplated in the strategy. Organisational performance ultimately would depend on 3 core elements outlined below:

(i) Organisational environment/business climate.
(ii) Business strategy.
(iii) Organisational capability.

Organisational Environment

In any organisational environment, managers/employees are exposed to pressures exerted by different agencies:
- (i) Pressures from present and potential would-be competitors.
- (ii) Pressures arising out of certain regulations and readiness to fulfil the expectations of clients/consumers and staffers.
- (iii) Pressures arising out of changes in technology and other related factors.
- (iv) Pressures arising out of reactions of managers to possible environmental changes.

"Gestalt Approach" to Organisational Environment

The Gestalt Therapy invented by Dr. F S Fitz, has been the focus of attention in the approach adopted by S M Herman. Here too much accent is laid on the natural development of the individual wherein he should be given adequate help to recognise, develop, as well as experience his potency and coping capability to deal with the organisational environment.

In this regard, the following approach has been suggested;
- (i) The individual should be fully encouraged to recognise his position/status and the 'behavioural changes' expected from the individual should be suitably encouraged.
- (ii) Fear and other inhibitions, which act as road blocks to individual expression in an organisational setting should not be the constraints.
- (iii) The culture of 'authoritarianism' normally found among managers controlling a team of employees, should be replaced by frank and free discussions without the syndrome of omnipotence. The boss and the subordinate should have better *'inter personal relations'*, and develop a richer understanding of each other so as to coordinate

and create a better environment to work together fulfilling the 'psychic needs'.

(iv) The Gestalt Approach induces transactions (interactions) among people at all levels which are authentic, much healthier and durable in effect.

Need for Congenial Work Environment

Only a congenial work environment would induce individuals (staffers/employees) to put forth their best performance besides helping to enhance their 'self-esteem'

An organisation, functioning in a highly regimented environment with little or no two way communication will damage the individual ability for creative problem solving.

Power hierarchy would increase mistrust among employees, and with mechanical routine ways of functioning would ultimately pave the way for poor decision-making.

The 'pyramidal values' of the 20th century business milieu may of course help to some extent the organisation's objective to achieve certain short-term goals! To make the long-term goals more effective, and to achieve organisational excellence, it is utmost essential for the modern CEO to discard controls, rewards and punishments. On the other hand, 'effective harmonious human relationships' should constitute the backbone of the aspiring modern firm to make it more vibrant, lasting and durable. New ways and methods, processes and values should be devised so that employees acquire the capability to tackle challenging situations.

Business Strategy

Ansoff, known as the father of the concept of 'corporate strategy' formulated some wonderfully practical ideas in regard to the role of 'strategic decision-making', which hold good even in the 21st century milieu.

He identified 4 decision types on which an healthy organisation should concentrate:
- *(i)* Decisions regarding strategy
- *(ii)* Decisions relating to policy matters
- *(iii)* Decision making on programmes
- *(iv)* Decisions on standard operating procedures

Ansoff indicated that the last 3 decisions were almost similar in nature, since they were designed to resolve recurring issues and problems without requiring significant time (or duration) for their management. On the other hand, he indicated that 'strategic decisions' had to be applied, and should be well-attuned to latest contingencies and should therefore be formulated freshly each time.

The 'Rover Group' of United Kingdom has made an excellent review of its foremost objectives outlining 'managerial principles' which would govern its organisational culture.

Figure 5.1: Vision Mission and Rover Group's Success Factors

Vision: Success through 'People': Extraordinarily customer's satisfaction
Mission: To develop and implement 'People Strategies' and 'Plans' to enable the Company to achieve 'objectives'
Critical Success Factors: 1. Create the culture 2. Help the leaders lead 3. Achieve "World Class" resourcing standards 4. Create continuous learning 5. Ensure Company wide understanding of the compelling business needs. 6. "Empower" individuals and teams.

Its 'Vision', 'Mission' and 'Critical Factors' furnished in figure 5.1 serve to highlight the principles meticulously followed by this renowned firm.

Management wizard Tom Peters has the firm conviction that in order to be successful in the 21st century business ambience,

the modern firm should have unconditional veneration for the following tenets:
- (i) Few layers ('flat') of organisational structure
- (ii) Autonomous-modules (independent units)
- (iii) Taking up manufacture of high value-added products and services with accent on 'product differentiation' (uniqueness)
- (iv) Creation of 'niche markets'
- (v) Adopting quality and service conscious approach.
- (vi) Responsive to environmental changes.

Several management experts (notably Dr Guest) insist on 'optimal employee commitment' to 'enterprise-goals and practices' for reaping substantial rewards.

Lean Production

The lean production technique, pioneered by Toyota Motor Company of Japan has won worldwide recognition. It describes a combination of techniques which facilitates firms to attain low cost status. The ideas of Dr W Edwards Deming has much influenced this line of approach. It is mainly based on 3 principles.

1. The first principle relates to 'just in time' production. Toyota felt that there was no point in producing cars (or any product) in blind anticipation of demand by customers. Waste must be thoroughly avoided. Besides, production should take place in consonance with actual market requirements.
2. Responsibility for ensuring quality rests with every one in the organisation and any quality defects perceived should be rectified immediately after identification.
3. Recognition of the 'value stream' concept is quite elusive. Instead of looking at the firm as a 'series of unrelated products and processes', it should be seen as a

"continuous and uniform whole" – a stream which includes suppliers of machinery, raw materials, components as well as customers.

Toyota of Japan launched its sophisticated 'Lexus Car' in 1990, known for its high performance – standards and luxury, and which surpassed in several respects even Mercedes and BMW. Toyota came to the forefront only after the introduction of 'Lexus' and continued to stay ahead for several years. Indeed Toyota had acquired unrivalled reputation by following Dr Deming's 'quality gospel', and efficiently and effectively exploited opportunities in the car market.

Other Advantages of Lean Production

(i) Womack and Jones claim that lean production practices would spread to all the types of manufacturing in due course. They observe:

"The adoption of lean production, as it inevitably spreads beyond the auto industry, will change everything almost in every industry. It will supplant both 'mass production' and the remaining outposts of 'craft production' in all areas of industrial endeavour to become the 'standard global production system' of the 21st century."

(ii) Dr Dankbarr believes, "It makes 'optimal use of the skills' of the work force, by giving workers more than one task (Multi-skilling) by integrating direct and indirect work and by encouraging 'continuous improvement activities' (quality circles). As a result, lean production is able to manufacture a 'larger variety of products' at 'lower costs' and 'higher quality' with less of every input, in, comparison with 'traditional mass production involving less human effort, less space, less investment and less development time."

Building Organisational Capability
World Class Operations-Role of Quality

Almost a decade ago, quality emerged as a strategic factor in several western firms owing to following reasons:

(i) A large number of new entrants with capabilities enhanced competition between new and existing companies, all of whom must compete to world class standards.

(ii) Customers have a wider variety of choice mainly on account of new entrants.

Since many firms have become over familiar with quality in recent years, quality guru Hamel has remarked, "The challenge is no longer quality; nor is it globalisation; you have been there, done that, got the ISO: 9001".

However, continuous improvement has become an integral part of quality and no firm can afford to relax just because it has achieved 'quality certification'. Besides, world class quality is not associated only with high end market tastes. Since each market segment may differ considerably in regard to the needs and requirements, the aspiring firm should meticulously identify these requirements (segment-wise) and provide customer satisfaction through world class operation capabilities.

In certain product lines, just meeting customer satisfaction would be the minimum requirement. Here the over-riding need to provide 'customer delight', instead of customer satisfaction assumes greater significance.

Management guru: Tom Peters has explained this as *going beyond customer satisfaction to customer success*. The following Figure 5.2 illustrates the emergence of 'world class quality' as presented by: Steve Brown, Blackmon, Cousins and Maylor.

For achieving world class quality, the following important factors should be given greater accent:

Organisational Changes

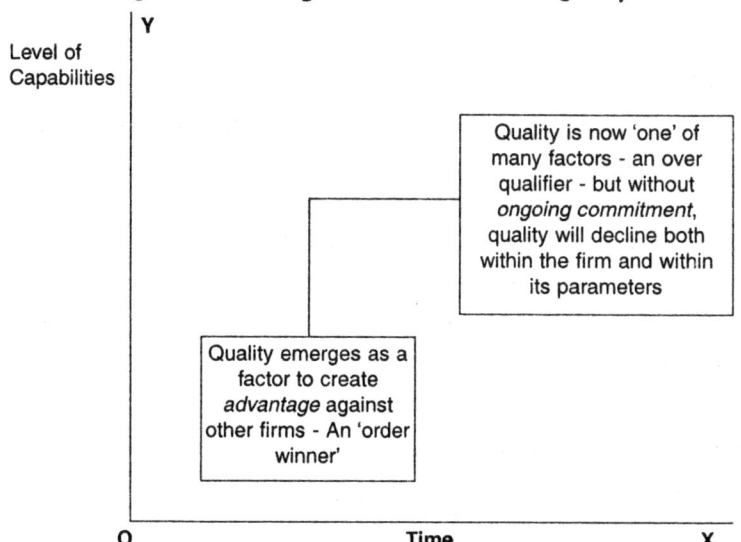

Figure 5.2: Emergence of World Class Quality

Top Management Commitment

Besides setting an example the firm should sensibly invest in training and other important features of TQM. Here, Dr Juran's essential idea on providing 'quality revolution' provides much insight into the process of spreading the message (among Japanese firms).

Accordingly, the senior executives of Japanese companies took personal charge of managing for quality. The executives trained their entire managerial hierarchies in how to 'manage for quality'.

Continuous Improvement

In spite of different viewpoints expressed by quality gurus: viz. Deming, Juran, Crosby and others in regard to actual approaches and prescriptions for quality, the most significant factor that emerges from these discussions pertains to the idea that 'quality as such is a moving target', and therefore the aspiring firm should have a strategic commitment to improve performance on a constant basis.

Quality Culture

Quality would encompass all personnel within the firm, as well as outside aspects such as supply chain and strategic partnership. Dr. Deming has categorically stated that at the outset "managers should drive out fear" from the work place, in order to achieve anything worthwhile. Freedom or openness is the key to quality and innovation and as such it should pervade all kinds of relationships and various networks with which the organisation is closely involved.

Total Quality Management

Total Quality Management should never be treated as a quick fix solution but should become an everlasting approach in managing quality standards.

In the turbulent and highly competitive milieu, a firm which has achieved success should not rest upon its laurels earned. It should have the steady commitment to improvement and innovation on a continual basis in order to emerge as a leading firm.

World Class Operations – Role of Innovations

Management wizard Tom Peters has succinctly observed, "Get innovative or get dead". Prof. John Kay has categorically stated, "Innovation is one of the very foundations of corporate success". In this regard, Gary Hamel has stated: "Never has the world been a better place for 'industrial revolutionaries' and never has it been a more dangerous place for complacent incumbents!"

In the present business scenario, being the first entrant to the market and having 'world class innovation capabilities' alone will not ensure sustained success.

In several cases the firms which emanate as followers have outperformed leading firms who were in the forefront to enter the market place.

Organisational Changes

There is no dearth of examples of firms which had failed to develop next generation products have been outperformed by their rivals-viz., market followers having world class capabilities. The following table 5.1 provides a list of few such firms which were overtaken by rival firms who offered better products.

Table 5.1: Firms overtaken by rivals

Sl.No.	Name	Product Manufactured
1	Ampex	VCRs
2	Xerox	PCs and plain paper copiers
3	Emi	X-ray scanners
4	Bowmar	Pocket calculators

The following table 5.2 gives a list of leading firms, which besides adopting world class operators have been constantly reinventing themselves from the very conceptual stage.

Table 5.2: Firms which had reinvented

Sl.No.	Name	Product Manufactured
1	Procter and Gamble	Disposable-diapers/consumer items
2	Pilkington	Float glass
3	Polaroid	Instant cameras
4	Corning	Fibre optic cable

The above named firms have stayed as market leaders on account of their superior operational capabilities of innovations.

As the distinguished economist, Prof. Joseph A. Schumpeter had stated that 'innovation' could also lead to 'creative destruction' of an entire industry sector in which the existing firms would be replaced by more capable new entrants with new products, new technologies, and highly responsive organisations.

World Class Operations: Capability to Innovate

The 'innovative capability' of an organisation is the most important attribute which is instrumental in transforming knowledge into competitive advantage.

Especially in sectors such as: pharmaceuticals, electronics, chemicals, aerospace and so on, the significant and critical mass of innovative capabilities essentially emanate from the research and development activities carried out by the leading firms in these sectors.

As such any aspiring organisation should make its utmost endeavour to track new technological developments taking place, in addition to maintaining capabilities in its core technology areas.

Intellectual Capital

Thomas A Stewart of *Fortune* magazine has categorically stated that 'knowledge' has become the most important factor in economic life. It is the chief ingredient of what we buy and sell, the raw material with which we work. 'Intellectual capital', not natural resources, machinery or even financial capital has become the one indispensable asset of corporations.

He further adds "intellectual capital" is intellectual material – knowledge, information, intellectual-property, experience-that can 'be put to use to 'create wealth' ! It is 'collective brain power'.

Actually it is "intangible stuff" – ideas, imagination, and know-how, besides knowledge workers who are experts in software engineering, advertising – executives and management consultants, who really make worthwhile contributions.

In the new knowledge economy, Stewart says, the rise of 'knowledge worker' fundamentally alters the nature of work, and the agenda of management. 'Managers' are custodians who protect and care for the assets of a corporation; when the 'assets' are 'intellectual', the managers' job changes.

Stewart emphatically states that the "rise of knowledge workers means that the bosses no longer know more than the workers". On account of this logic of "management-pyramid", a small group of people on top telling a large number of employees "what to do" becomes redundant.

When knowledge economy becomes widespread, new jobs such as: (i) Chief Knowledge Officers, (ii) Directors of Intellectual Capital are being created in several modern firms.

When change occurs, several dimensions are affected:
- (i) Changes in the markets/industries
- (ii) New methods of competition
- (iii) Totally new entrants to the field
- (iv) Globalisation in markets/supply chains
- (v) Changes in technologies.

Ultimately, it results in 'product and process innovations' which are instrumental in considerably altering the cost structures.

When continuous regeneration and development of organisational knowledge takes place, it becomes imperative for the organisations and the staffers associated to prepare for continual learning.

The need to continually reinvent the modern organisation through constant learning would become the key ingredient and essential feature of 'knowledge management' in the 21st century.

The organisational culture should ardently support the ability to create, absorb and assimilate new knowledge', besides abandoning outmoded knowledge procedures and techniques.

Indeed in several product lines competitive advantage mostly occurs only through innovations – whether in the realm of products, processes or services.

Several modern firms take decisions 'on investment' on the assumption that they have the required intellectual capabilities to create 'knowledge' that will lead to 'innovations' and eventually to greater profits.

APPROACHES TO IMPROVEMENTS

(A) Breakthrough Improvements
(B) Continuous Improvements

A. Breakthrough Improvements

Breakthrough improvement is actually 'innovation based improvement', wherein major and dramatic changes take place in course of operation leading to faster and enhanced performance.

The following examples may serve to highlight this concept:
   (i) Installing a new more efficient machine in a factory.
   (ii) Introduction of a new and sophisticated degree programme at a university.

As a result of such improvements an organisation goes a step ahead and there would be perceptible changes in performance standards.

Though such improvements may be quite expensive with substantial investment of capital, they may thoroughly change ongoing operations involving modifications in product, services offered as well as process technology.

The following figure 5.3 serves to illustrate the actual results obtained as a result of 'breakthrough improvements'.

Figure 5.3: Actual Performance Improvement Obtained in Relation to Intended (Breakthrough) Improvement

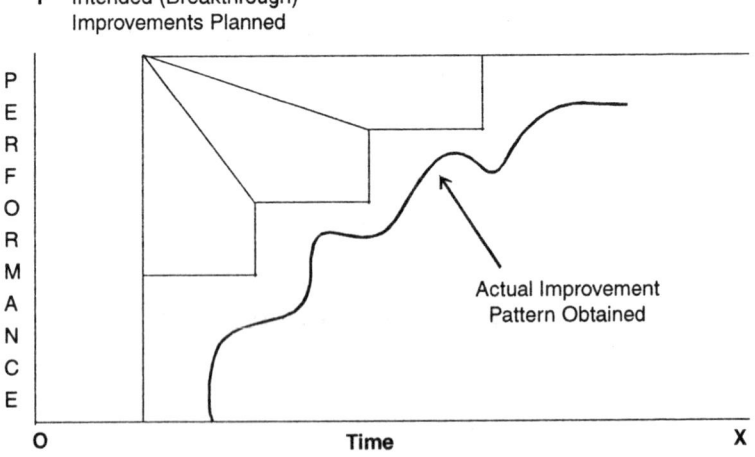

B. Continuous Improvements

In respect of 'continuous improvement', a firm opts for an approach which results in increase of its performance standards in small incremental steps as indicated below:

- (i) Reducing change overtime in manufacturing a product.
- (ii) Making minor modifications in design/delivery.
- (iii) Reducing work load for students in an university course by rescheduling its assignments.

These improvements could be undertaken relatively painlessly along with all other incremental improvements. In continuous improvement, the momentum of improvement assumes great importance. Even if successive improvements are small, the most important factor is that the firm should be keen on undertaking monthly, weekly and quarterly improvements in a steady fashion.

In this connection, reference may be made to 'Kaizen', a Japanese term for continuous improvement, first propounded by Masaaki Imai in Japan which also encompasses improvements in personal life, home life, social life and work life.

When applied to the work place, the critical issues concerning production like quality, cost, delivery, safety and morale etc., in regard to product or process are identified. After identifying certain critical issues, it is easy to enlist the co-operation of all managers and staffers.

Japanese experts believe that the ability to improve on a continuous basis is not something which always comes naturally to managers/staffers. In order to sustain continuous improvements over the long-term, it is felt that the specific abilities, behaviours and actions have to be consciously developed.

The pattern of performance, improvement, and continuous improvement (the 'Kaizen' approach) being adopted by a modern firm is highlighted in figure 5.4.

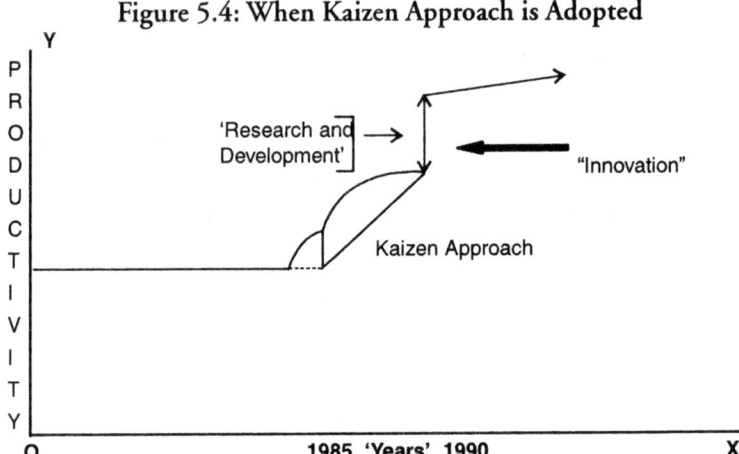

Figure 5.4: When Kaizen Approach is Adopted

In a modern firm adopting the Kaizen approach at the outset, a model area would be selected for undertaking a trial project. If tangible results are achieved, the methodology of approach would be finalised with suitable standardisation. Later expansion takes place to upgrade model areas. Kaizen projects are taken up for a duration of 3 to 6 months normally. The company-wide implementation may take roughly 4 to 5 years in respect of a large organisation.

It is clear that breakthrough improvements offer high value creative solutions by going to first principles. Such breakthrough improvements are a series of dramatic explosive changes. While breakthrough improvements could be compared to highly impressive sprints requiring great expertise in the field of sports, continuous improvements are similar to marathon running which keeps the firm going.

The essential features of both approaches as highlighted by Imai are furnished in figure 5.5.

Figure 5.5: Essential Features of Breakthrough and Continuous Improvements

Sl. No.	'Feature'	Breakthrough Improvement	Continuous Improvement
1	Effect	Short term; But dramatic	Long term, Long lasting; undramatic
2	Pace	Very big steps	Small steps
3	Time Frame	Intermittent	Continuous/Incremental
4	Approach	More on individual ideas/Contributions	Collective-Group efforts; Systems Approach
5	Orientation	Technology	People
6	Evaluation	Results for substantial profits	Process/Efforts for better results

Achieving Competitive Advantage and Success

Management wizard Michael Porter provides certain wonderfully fine tools and techniques to earn "supernormal profits" by exploiting certain competitive advantages "within an industry framework".

He is of the opinion that competitors continually jostle for greater customer attention by certain methods:

(i) By adding more features or extra services at no charge
(ii) By cutting prices
(iii) By supplying extra product at the same price.

Several airline advertisements bear ample testimony to the above. They claim to provide more comfortable seating arrangements with greater coverage of destinations than their competitors.

In this context, Michael Porter has suggested certain "generic strategies" as outlined below:

Cost Leadership

The cost leadership strategy involves being the lowest cost producer within a broad industrial sector. Here an organisation is using various sources of cost advantage, besides quoting prices at a level that is comparable with or lower than its rival firms. By considerably increasing the sales volume, it can achieve higher gross margins (profits). It is utmost important that the low cost producer must ensure a level of functional performance of (his offerings) and quality standards found acceptable by the market.

Differentiation

Here a particular firm tries to achieve an unique position in the industry by concentrating on certain dimensions much valued by the consumers. It carefully selects 2 or 3 attributes which the buyers perceive as most important and uniquely positions itself for getting a premium price for its products having 'unique excellence'.

Focus

A modern firm has the options to pursue advantages from 'cost leadership' as well as from 'differentiation'. It can adopt a strategy of focus-wherein certain segments as groups are selected exclusively either for 'cost strategy' or 'differentiation strategy'

By optimising the strategy for the target segments the firm focuses greater attention to achieve competitive advantage. The firm must have clear notions on consumer preferences, geographic coverages, product applications, distribution channels, etc., with due regard for age-group, gender ethnicity, income-brackets, etc.

The following figure 5.6 provides a clear picture of the strategies suggested by Michael Porter.

Figure 5.6: Competitive Advantage

		Lower Cost	Differentiation
Competitive scope	Broad target	Cost leadership	Broad differentiation
	Narrow target	Cost focus	Differentiation focus

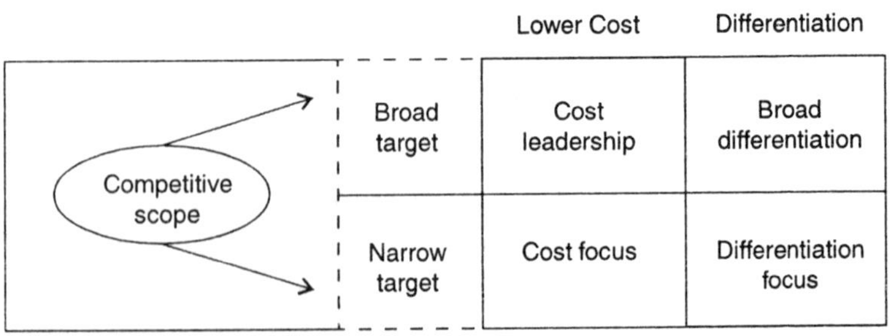

Other Important Factors

Efficiency in the workplace would go up with the capacity to innovate.

An organisation would be successful if its employees are creative either by improving the existing products or by innovating new products.

Most of the technology based companies are science based, and new ideas in the form of inventions are generated by the R AND D laboratories attached to them. But these inventions become 'innovations' only when these are transformed into usable products.

'Technological innovation' as such would encompass:
- *(i)* Creation of new knowledge.
- *(ii)* Generation of useful technical ideas which would ultimately enhance the value of the product and improve the features so as to attract consumers.
- *(iii)* It covers improving the manufacturing processes as well as services.
- *(iv)* Initially developing suitable proto-types and subsequently taking up manufacture, establishing 'distribution channels' and so on.

The products should reach the target audience in consonance with their expectations. Successful technological innovations emanate from experts having a combination of entrepreneurial, managerial and technological skills.

Companies which are keen on remaining in the forefront should increase their *competitive advantage* by several techniques such as:
- *(i)* By decreasing the cost of production
- *(ii)* By improving quality
- *(iii)* And by rapidly responding to customer needs.

The management must initiate a highly conducive climate which accentuates creativity and innovation. Employees should be permitted to adopt innovative approaches and techniques by setting up appropriate goals and objectives.

Innovations should be suitably rewarded and an aspiring employee should be assisted to develop himself with appropriate education/training as well as exposure to current developments.

To achieve success and to sustain in the technologically competitive market the modern firm should establish a strong information system to gauge the position enjoyed by its competitors.

The 21st century extends a warm and cordial welcome to all organisations who are keen on leveraging talents which would ultimately create wealth. In the current milieu, only firms which could raise the productivity of knowledge workers would become market leaders.

In creativity, the enthusiasm and emotional component emanating from the right brain patterns occupy the pride of place. The rational left brain activities responsible for logical thinking/reasoning are mostly supplemented. Since creativity is the foremost requisite for wealth creation, it is of crucial importance to provide the right environment and also ensure that the structure, culture, people and policies lend all possible support for creative efforts.

The dynamic global firm should formulate new strategies having core processes far superior to its competitors. The hallmark of success would ultimately depend on factors such as:

(i) New equipment installed
(ii) New standards for quality imposed
(iii) Improved methods of training taken up
(iv) Innovative methods for effective distribution

The following figure 5.7 entitled 'strategic staircase' outlines the series of initiatives to be taken up by the global firm to 'broaden the capabilities of the organisation as a whole' over a period of time.

Organisational Changes

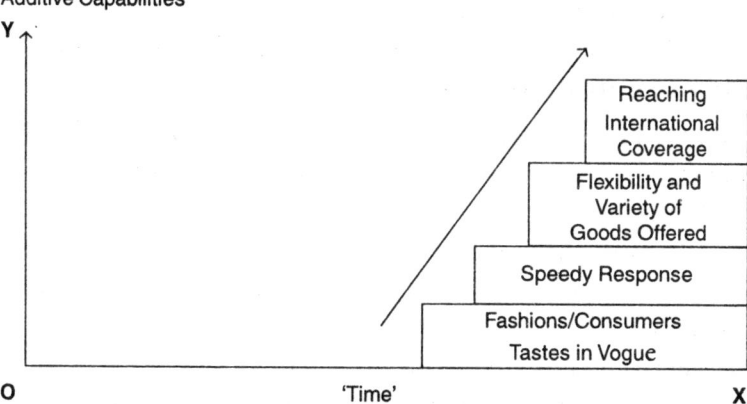

Figure 5.7: Strategic Staircase

Process Management Tool Box

As such there are different management tools and fads, which could be advantageously selected and used in specific situations. The tool box shown in figure 5.8 should not be indiscriminately used by the modern firm. It should select suitable tools and

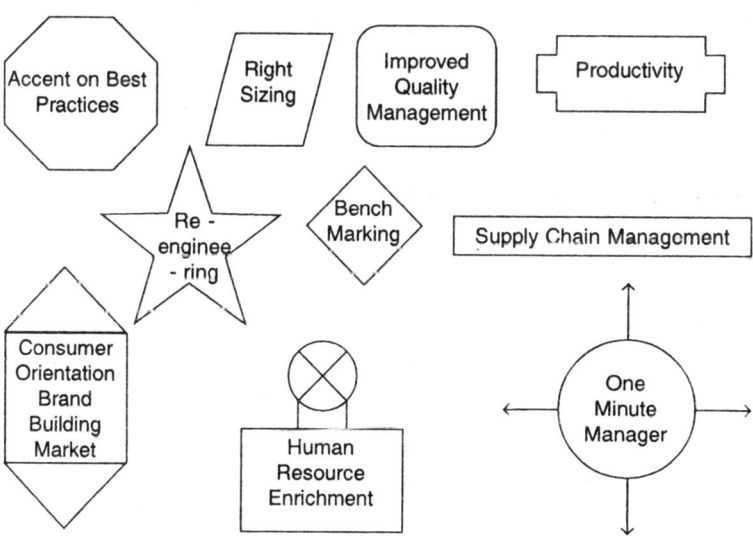

Figure 5.8: Process Management Tool Box

techniques out of the enumerated options keeping in view the long-term goals and objectives to be achieved.

The global firm should also give due regard for human values.

Purpose of Business

Charles Handy, the management expert, has provided an authoritative conclusion in regard to the 'purpose of business' on the following lines:

"The principal purpose of a company... is to make a 'profit' in order to continue to do things or make things, and to do so ever better, and more abundantly." He further says that profit is neither an end nor a means but an 'end product', the reward that flows from doing or making better and more abundantly what you want to do or make by 'optimising customer satisfaction'.

In this regard, it is interesting to note that Andy Grove developed an excellent 'business model' to a fine pitch of performance. According to him a good business model basically hinges on 'sales volume, realised prices, and gross margins.' Andy Grove had great sagacity in carving out a 'strategy' that kept competitors always out of his 'markets'. With great tenacity he consolidated Intel's power and position, and facilitated it to enjoy a dominant market share, by accelerating technology and by making micro-processors and personal computers-vastly more powerful and vastly cheaper.

Importance of Intangibles

Several most admired global firms attach considerable importance to 'intangibles' (such as innovation, high customer satisfaction etc.) instead of solely depending upon financial performance as the yardstick mostly used for measuring success. Only a highly reputed firm maps qualitative measures such as: innovations, trust, regard, team work and diversity etc., recognising that impressive achievements in these areas are also critical to its outstanding success in revenues and profits and other indicators of financial performance.

The Hay Group's study of the 'globally most admired companies' as reported in 'Fortune' magazine has presented the following conclusions:
- *(i)* Only companies setting more challenging goals and linking compensation more closely to the achievement of these goals have maintained enhanced reputation throughout.
- *(ii)* Their activities are oriented towards long-term performance and these companies rely on customer indicators like 'utmost satisfaction, loyalty, and market share.'
- *(iii)* Several such companies concentrate on career development and other employment oriented measures.

As such, it is extremely difficult to value an 'intangible'. Even a tiny flaw in a product or a component would simply damage a leading firm's great reputation.

Though the intangible reputation is quite tough to manage, it has to be cultivated over the years. It would certainly lend greater competitive edge. The following firms stand head and shoulders above the rest of their rivals on account of certain characteristics as outlined hereunder: (in table 5.3)

Table 5.3: Intangible Qualities

S.No.	Firm	Intangible Quality
1	Toyota Cars	Bullet Proof Quality
2	Dell Computers	Custom Designed Products
3	Intels	Technological Leadership
4	Gen. Electric Co.	Famed Management System

Ethical Practices

As such a modern business organisation could be built on certain strengths such as superior technology and talented personnel. But at the same time it should not be forgotten that 'high standards of product excellence, uncompromising integrity and lofty ethical standards' are the essential ingredients for reaching greater heights!

6

Quality as a Cohesive Policy for Achieving Excellence

"We are the Inspection Department, and our job is to look at these things after "they are made" and find the bad ones. 'Making it right' in the first place is the job in the Production Department."

—J.M. Juran
Washington Post

What Is Quality

On account of unprecedented competition on a global scale, 'quality' has emerged as a major strategic factor and every forward focusing firm developing innovative products and processes must capitalise on this ingredient in order to achieve sustained success.

Lawrence Bossidy, CEO of Allied Signals who is keen on revitalising "total quality efforts" has succinctly observed:

"Total quality: team concepts... six sigma... big bold initiatives are waves of 'rejuvenation' that roar through a company washing away routine and fired thinking. The whispering cynics of bygone days called them 'flavour of the month' or 'the latest hula-hoop'. But they were wrong. If pursued with passionate commitment, they renew and excite a company. But they must be refreshed and taken to 'new levels' on a regular basis."

Reasons for Transformation

Quite a few decades ago, 'world class manufacturing' was mostly restricted to industrialised countries of the world. In the present decade even developing countries, which were till now handling labour intensive production of simple products, have turned their attention towards the manufacture of most sophisticated products involving much precision. The revolution in information technology has brought about sweeping changes all over the globe. The ability to coordinate global activities among firms have considerably increased after the emergence of global IT architectures with common operating environments.

Though 'quality' is often used to signify 'excellence of a product or service', a modern firm, besides supplying products/services to specific standards, can hardly overlook the need to include in the assessment of quality, the true requirements of the customer–the needs as well as expectations.

The management philosophy which has gained universal recognition nowadays pertains to 'total quality management'. The term is quite comprehensive and covers concepts such as 'continuous process improvement', 'quality first', 'do it right the first time', 'kaizen', 'six sigma', 're-invention', 're-engineering' and so on.

Reliability also ranks in importance with quality since the acceptability or otherwise of a product or service will depend much upon its ability to function satisfactorily over a period of time. Only by consistently meeting the customer requirements the modern firm could hope to delight the customers, besides enlisting customer loyalty.

A sophisticated modern firm should be able to meet the customer requirements and its reputation would be built on certain principles enumerated in figure 6.1.

Figure 6.1: Principles Determining Quality/Performance-Standards

1.	Firms' sustained commitment to excellence
2.	Laying accent on long term view of the future
3.	Main focus on 'customer satisfaction' and ultimately achieving customer delight
4.	Excellent opportunities for: (a) Continuous learning (b) Product related learning-globally (c) Improvement-mentality
5.	Right organisational climate to encourage more of: (a) Employee-involvement. (b) Empowerment (c) Creativity
6.	Importance of most efficient team work by pooling of expertise/resources-which help people to tackle complex problems and grow.
7.	Excellent process and management with a 'defect prevention' philosophy.

New Standards and Strengths required in the 21st Century Milieu

1. Quality of product, process and service has become the greatest focus for the modern firm.

    'Quality' is considered the minimum entry point into global market. To compete globally firms should be able to offer sophisticated products of high quality and services. Leading firms which have an edge over others always focus on 'consumer delight' rather than on 'consumer satisfaction'.

2. With globalisation people are experiencing mind-boggling changes, and there are rapid changes in consumption patterns. Modern firms are confronted with world class quality competition and customers have acquired certain new characteristics;

    (a) They demand increasing variety of products

    (b) Buyers have become highly discerning and switch over to costlier items, depending upon their aspirations and value for money equations

3. Consumers increasingly demand customized products to realise their desires for self-fulfilment. Nowadays several modern firms acknowledge one of the best ways of providing service is to ask the customer what he wants, and then give it to him. The inventory system should, therefore, be suitably streamlined to provide good customer service.
4. It is not always easy to capture substantial market share solely by manufacturing quality products. The renowned 'Sony' by incorporating 'portability' and 'miniaturisation' as core competencies started manufacturing portable radios and tape recorders. Thereby they enhanced their market share substantially.
5. In the global scenario customers are demanding that the highest quality product should be made available at the lowest possible price. Reducing costs in the production processes has gained greater momentum in the present decade.
6. As such new demands for quality consumer products, changing tastes, fashions and fads at the global level and short product cycles have started pushing modern industrial enterprises to seek worthwhile global partnerships so as to gain access to the emerging new markets.
7. Tom Peters has stated that in most of the leading modern firms the top executives are much obsessed with significant quality improvements, than they were in previous decades. Quality improvement programmes have received greater accent especially in IBM, Texas Instruments, Milliken and Co., Ford, Air Force Tactical Air Command, and the executives assign the pride of place to quality improvements rather than for financial reviews in monthly marketing agendas.

8. In the western manufacturing firms, the top management commitment for achieving phenomenal success is so great that no one can walk past a shoddy product or service without comment and appropriate action within the factory premises.
9. Quality is recognised and rewarded in firms like Ford. Quality targets have become part of 'Ford's consecutive compensation plan'.
   During 1985-90, for Ford managers approximately 40 to 65% of their bonus was mainly based on contributions relating to quality and only 20% was assigned for contributions towards greater profit.
10. Effective training in technologies meant for assessing quality has received greater attention in western countries as indicated below:
    (a) Everyone has to be instructed on problem, cause and effect analysis, rudimentary statistical-process-control, group problem solving and interaction techniques.
    (b) Since world class manufacturing companies of Japan and Germany spend prodigious sums of money on training at present, the western firms have started working on the same fashion and they also believe the Japanese axiom 'quality control starts in the training and ends in training'.
11. The leading western firms believe in 'Hawthorne Effects' in the work environment to improve productivity/quality. They set up new goals, new themes, new rewards, new team champions, new team configurations, new events to celebrate achievements, etc. However, the basic system remains the same. They organise 'zero defect programmes' periodically.
12. It has become customary in the western world to include leading suppliers, distributors, and customers as part of

the organisation's quality improvement, studying reasons for the phenomenal achievements of some firms for example: Toyota, Ford's taurus project, Hewlett Packard etc., are undertaken. They are keen on making every product getting relatively better (or even worse) but it should not be static (or stand still).

Ford is reported to gather approximately 4000 to 5000 new suggestions and carry out experiments to decide upon successful implementation.

Quality Management: Viewpoints of Leading Architects

Different approaches were suggested for quality management by the internationally renowned... gurus, and the main focus of their teachings could be summarised as below:

Quality Guru	Main Focus
(i) Crosby	Company Wide Motivation
(ii) Deming	Statistical Process Control
(iii) Feigenbaum	Systems Management
(iv) Juran	Project Management.

Crosby's Approach

In Crosby's programme the main audience is 'top management".

He is much inclined to stress the need for achieving increasing profitability through effective quality improvements. He expresses the view that high quality invariably 'reduces cost and raises profit'. The five salient 'absolutes' enumerated by Crosby are presented in figure 6.2.

Figure 6.2: Five Salient Absolutes of Crosby

- Quality is defined as conformance to requirements, not as "goodness" nor "elegance"
- There is no such thing as a quality problem
- It is always 'cheaper' to do it right first time
- The only performance measurement is the cost of quality
- The only performance standard is "zero defects"

Crosby, while defining 'quality', indicates that a quality product or service he is referring to pertains to one which meets the requirements of the customer or user. This means in turn that those requirements must be defined in advance, and that 'measures must be taken continually to determine conformability.

The requirements may include both 'quantitative' and 'qualitative aspects', although Crosby's target emphasis is towards the 'quantitative' that is 'zero defects'. He is of the firm conviction that 'quality is essentially measurable' and it is achieved when 'expectations' or 'requirements' are met.

Crosby's second absolute explains that "there is no such thing as a quality problem". It could be stated that poor management creates the quality problems, and they do not create themselves or exist, as separate entities from the management process. Ultimately, the product and its quality do not exist in a vacuum, since they are the resultant of a management process, and if that is a quality process, then a quality product will emerge. Here the belief is expressed that management must lead the workers 'towards a quality outcome.'

Thirdly, Crosby states, "It is always cheaper to do it right first time". Logothetis has suggested that "A company which relies on 'mass inspection' of the final output to improve quality is doomed to stagnation".

Crosby clearly lays accent on his belief that 'inspection' is a cost and that 'quality needs to be designed into a product' – not that flaws should be inspected out. Here there is a strong belief in the 'potential to achieve quality' – that is conformability to requirements by developing a "quality process and product from the outset" with no expectation of failure. Here "prevention of error is considered better than rectification!"

Fourthly, Crosby has expressed his belief that the 'cost of quality' is always a 'measurable item' – as for example: rework, rejects, warranty cost and, that this is the only basis on which 'performance' could be measured. As Logothetis states that it is the 'price of non-conformance'.

Finally, Crosby argues that the only "performance standard is zero defects". Here, the idea is that "perfection is the standard to aim for" through "continuous improvement", and underpinning that "zero defects" is an achievable and measurable objective.

Here also Crosby's inherent belief in the quantitative approach to quality is made amply clear with 'perfection' that is "zero defects" suggested as the main target.

There are essentially three strands of thought in regard to Crosby's approach.

(i) An inherent belief in 'quantification'
(ii) Dependency on 'management leadership'
(iii) 'Prevention' is better than 'cure'

As such in his approach 20% of manufacturing cost is attributed to failure and the workers should not be blamed for errors, since 85% of all quality problems fall within the purview/control of top management.

Crosby has also formulated 'a quality vaccine' encompassing 21 areas, which are divided into 5 major categories – Integrity, Systems, Communications, Operations and Policies – which he considers as "preventive medicine" for "poor quality"! Crosby is a great motivator of senior management who are actually involved in initiating the improvement process.

Crosby's 14 Step "Quality Improvement Programme" has been displayed in figure 6.3.

Figure 6.3: Crosby's 14 Step Programme

Step	Action to be taken
Step (1)	**Establish Management Commitment** Whole management team to participate; and half-hearted attempts would fail.
Step (2)	**Constitute Quality Improvement Teams** Need for having a multi-disciplinary team is stressed. It should not be confined to quality department alone
Step (3)	**Establish Quality Measurements** Every 'activity of the company' to be covered; all aspects: design, manufacture, delivery, etc., to be covered with measurements to 'provide a platform to the next step'
Step (4)	**Evaluate Cost of Quality** Evaluation should make use of measures established in Previous step and to highlight where 'quality improvement' will be profitable
Step (5)	**Raise Quality Awareness** Training staffers through excellent communications with: books, videos, displays, posters, etc.
Step (6)	**Take Corrective Action** Encourage staffers to identify and rectify defects and intimate higher supervisory levels
Step (7)	**Zero Defects Programme** A separate committee to function for initiating and implementing a 'Zero Defects Programme'.
Step (8)	**Training Managers/Supervisors** Focus on achieving understanding by all managers/supervisors... of different steps in quality programme and they should be able to teach others
Step (9)	**Organise a Zero Defect Day** To create awareness on the 'expectations' of the firm-in a celebratory atmosphere
Step (10)	**Encourage Goal Setting** To make rapid improvements: Encourage 'goal setting' with suitable 'time frames'
Step (11)	**Reporting: Obstacles/Errors** Employees should be free to inform management on factors which act as hindrance to error free-work; cover defective/inadequate equipment, poor quality components, etc.
Step (12)	**Recognition** There should be a non-monetary reward scheme for those who make worthwhile contributions (giving shield/certificate/trophy etc.)
Step (13)	**Set up Quality Councils** Forums of quality-professionals + team leaders to discuss and formulate 'action plans' for greater quality improvements
Step (14)	**Do it all over Again** Since achievement of quality is an ongoing process, there is always further scope to improve

Quality as a Cohesive Policy for Achieving Excellence

To Sum up: Philip Crosby's Philosophy

Under the quality management philosophy of 'zero defects', the ultimate goal of operations is 100% conformity.

In part this is based on his idea that 'quality is free', and in this regard he argues that the 'benefits from improving quality' more than compensate their costs:

Philip Crosby convincingly states:

"Quality is free! It is not a gift, but it is free. What costs money are the unquality things – all the actions that involve 'not doing jobs right' the first time!"

Crosby spoke of '5 stages of development' and in the first stage the cost of quality may be around 20% of sales. Firms become enlightened over the years. Therefore by the fifth stage, the final stage the cost of quality would be around 2.5% (negligible).

W. Edward Deming's Approach

Dr Deming is known as the founding father of 'quality movement'. He is regarded as the 'chief architect' of the phenomenal industrial success achieved by Japan. Only subsequently the US industry (the home country of Deming) woke up to Deming's remarkable theories/contributions in the field of 'statistical quality control' and assigned him the pride of place.

Deming defines 'quality' in terms of 'quality of design', 'quality of conformance' and 'quality of the sales and service function.'

His foremost aim is to improve 'quality' and 'productivity' as well as 'jobs', so as to ensure the 'long-term survival of the firm' and help to improve the 'competitive position.'

Deming keenly advocates measurement of 'quality' by 'direct statistical measures' of manufacturing performance against specifications. His main theme indicates:

"While all production processes exhibit variation, the 'goal' of quality improvement is to reduce 'variation'".

Deming's methodology of approach has a high 'statistical slant', and he believes that every employee should be suitably trained in 'statistical quality techniques'. He is well-renowned for his systematic approach to "problem solving". The *Deming Shewhart* or 'PDCA Cycle': 'Plan, Do, Check, Action' is illustrated in figure 6.4.

Figure 6.4: PDCA Cycle

Two salient features emerge from the above:
(i) Here the quality initiative is 'in a systematic methodical approach' contrasting sharply with the ad-hoc and random behaviour of several quality initiatives.
(ii) Deming stresses the need for 'continuous quality improvement action'. It is in sharp contrast with the overtones in Crosby's approach which suggests a 'discrete set of activities'.

Later, while focusing attention on American and Western management practices, he wanted to 'eliminate 7 deadly sins' about bad management practices found in them in order to support the implementation of a 'truly successful quality initiative.'

Details in regard to these 7 deadly sins are presented in figure 6.5.

Figure 6.5: The Seven Deadly Sins

S.No.	Sins	Strands
1	Sin-1	Lack of constancy of purpose
2	Sin-2	Emphasis on short-term profits – forgetting long-term orientation
3	Sin-3	Poor performance appraisal-leading to bitterness, bruised-feeling demolishing team work
4	Sin-4	Job hopping – destroying team work/poor decision making forgetting healthy growth of organisation.
5	Sin-5	Running a company on "visible figures" only, when intangible effects' are forgotten.
6	Sin-6	Excessive medical costs met by the employer
7	Sin-7	Excessive costs of liability • Blaming a single person for fault when there is collective-effort.

As such Deming expects the managers to change, to keenly develop a partnership with those at the operating level of the business and to manage quality with 'direct statistical measures' and without 'cost of quality measures'.

He lays accent on the need of management to change the organisational culture, which is closely aligned with the Japanese approach.

Statistical Process Control

Deming has advocated another approach viz. 'Statistical Process Control', which is a 'quantitative approach' based on 'measurement of process performance.'

The quintessence of his argument is that 'a process is considered to be under control', which is stable throughout when its 'random variations' stay within the 'determined upper and lower limits'. In such a process all the special causes of failure are eradicated.

In the control chart presented as figure 6.6, the value of a measurement associated with an event is a 'process'. 'Statistical analysis' of the values recorded will help to reveal the mean value.

For a particular process the normal variation from this mean value is conventionally taken as any value with ± 3; standard deviation of the mean.

Figure 6.6: A Sample Control Chart

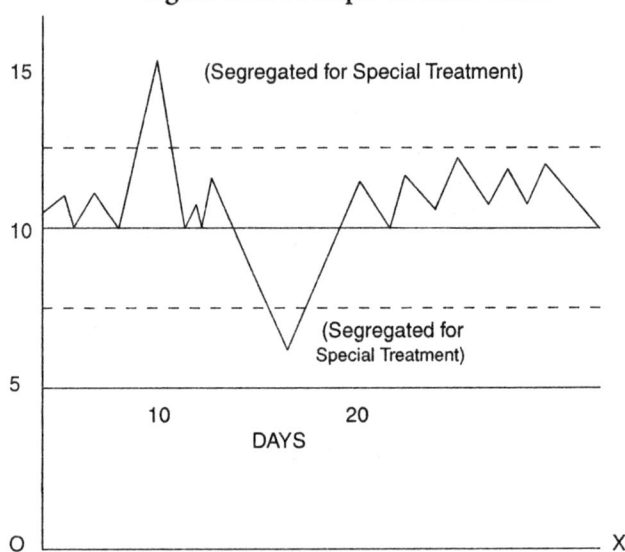

Events which fall outside that normal variation are termed as 'special' and should prove tractable to individual diagnosis and appropriate treatment. Events falling within the norms are treated as products of the organisation and of the system and, therefore, any treatment required (rectification) could be tackled at the usual management level.

It could be fairly estimated that possibilities for improvements are normally tackled in the following proportions:

- (a) 94% → belong to the system and could be effectively tackled by management
- (b) 6% → which fall outside the mean value require specialised attention.

Quality as a Cohesive Policy for Achieving Excellence

Figure 6.7: Deming's Principles for Transformation

Deming laid accent on continuous learning and improvement and favoured development of a quality culture by adopting the 14 Principles.

Principle	Deming's Main Philosophy
Principle-(1)	Create constancy of purpose for improvement of 'product' and 'service', with the aims being to become 'competitive and stay in business' and to provide jobs.
Principle-(2)	Adopt the new philosophy; Management must awaken to the challenges of a new economic age – learn their responsibilities and take on leadership for change.
Principle-(3)	Cease dependence on 'mass inspection" to create quality. "Build quality into the product" at the outset.
Principle-(4)	End the practice of awarding business on 'price tag' alone. Instead try to minimise total cost. Move towards long-term relationships of loyalty and trust.
Principle-(5)	Aim for continuous improvement of the system of production and service to improve 'productivity' and 'quality' and to 'decrease costs'.
Principle-(6)	Institute training on the job; improve 'competencies and skills' with objective of continuous improvement.
Principle-(7)	Institute leadership with the aim of supervising people to help them in a cultured manner.
Principle-(8)	'Drive out fear', so that everyone may work effectively
Principle-(9)	Break down barriers between departments. Encourage research, design, sales and production to work together, to foresee difficulties in production while initiating quick remedial measures.
Principle-(10)	Eliminate slogans, exhortations and targets for the work force. Such exhortations only create adversarial relationships.
Principle-(11)	Eliminate adherence to numerical goals!
Principle-(12)	Remove all 'barrlers' to 'pride of workmanship'.
Principle-(13)	Institute a vigorous programme of 'education' and 'self-improvement'. When people improve with greater knowledge, 'competitive advantage' will improve.
Principle-(14)	Put everyone in the company to work to accomplish the transformation; the 'transformation' is everyone's job!

Feigenbaum's Contribution

Feigenbaum, the renowned originator of the highly popular concept, 'total quality control' offers a precise and comprehensive enunciation:

"Total Quality Control is an effective system for integrating the quality development, quality maintenance, and quality improvement efforts of the various groups in an organisation, so as to enable marketing, engineering, production and service, at the 'most economical levels which allow for full customer satisfaction."

Fiegenbaum is of the opinion that 'quality' is a way of managing a modern business organisation. He has expressed the firm conviction that every individual in an organisation inclusive of the work force should actively participate, in order to bring about significant improvements. This would involve a thorough understanding by every one of the customer oriented quality management process for which the firm stands committed.

In order to achieve this, sometimes short-term motivational programmes of an organisation may have to be abandoned.

He further indicates that substantial financial rewards should not be forgotten. He believes that a highly effective 'quality management programme' should help to assure a 'best return on investment' for any specific project in the highly competitive environment.

He lays accent on 'quality leadership' which is a fundamental prerequisite for outstanding success in the market place.

The 'quality costs' have been segregated into 3 categories by Feigenbaum as given below:

(i) Appraisal Costs
(ii) Prevention Costs
(iii) Failure Costs

Feigenbaum states that: Appraisal Costs + Preventive Costs + Failure Costs = *Total Quality Cost.*

Feigenbaum has further stated that a modern firm should endeavour to reduce the 'total quality cost' which may centre on 20-30% of 'sales' or 'cost of operations', to still lower percentages and such 'endeavours should be an ongoing process within the aspiring firm.

Feigenbaum is of the opinion that phenomenal success depends ultimately on the managerial know-how, and commitment and the following factors are given much importance:

(i) Genuine endeavours to strengthen Principle-(4) the 'quality control process' itself.
(ii) To ensure that 'quality improvement effort' becomes the ingrained habit.
(iii) Managing quality and cost as complementary objectives.

For achieving 'Total Quality Success', Feigenbaum provides a set of 10 wonderfully fine benchmarks as shown in figure 6.8.

Figure 6.8: Ten Benchmarks Essential to Achieve Total Quality Success

1.	Quality is a company-wide process.
2.	Quality is what the customer says " It is ".
3.	Quality and cost are a 'sum'; not a difference.
4.	Quality requires both 'individual' and 'team efforts' [with great zeal].
5.	Quality is the way of 'managing'.
6.	Quality and innovation are mutually dependent.
7.	Quality is an 'ethic'.
8.	Quality requires 'continuous improvement'.
9.	Quality is the most cost-effective and least capital-intensive 'route' to 'productivity'.
10.	Quality is implemented with a total system connected with customers and suppliers.

Joseph Juran's Approach

Joseph Juran has made spectacular contributions in the field of 'quality control' than any other professional.

Juran's foremost contribution to 'quality management' is his methodology for determining the avoidable and unavoidable costs of 'quality' thereby providing a yardstick for measuring the quality programme.

Like Deming, Joseph Juran had also expressed much confidence in the development of 'quality management' in multifarious Japanese enterprises. Juran had the firm conviction that firms keen on achieving success should make every possible endeavour to reduce the cost of quality at the outset.

Juran's definition of quality provides a new strand of thought; he defines 'quality' as 'fitness for use or purpose.' Certain management experts have stated that Juran's definition is a more useful one than 'conformance to specification', since a dangerous product could conform to all specifications but still be unfit for use.

The most useful strand of Juran's thinking pertains to his 'Trilogy of Quality Planning', Quality Control, and 'Quality Improvement'.

The following figure 6.9 outlines the "Trilogy of Quality" formulated by Juran.

Figure 6.9: Quality Trilogy of Joseph Juran

1.	*Quality Planning:* Determine quality goals; implementation planning, resource planning, express 'goals' in quality terms; create the quality plan.
2.	*Quality Control:* Monitoring performance, compare objectives with achievements; action to reduce 'gap'.
3.	*Quality Improvement:* Reduce waste; enhance logistics; improve employee-morale; improve profitability; satisfy customers.

Juran ultimately lays accent on 3 areas:

(i) Changing 'management behaviour' through quality measures

(ii) Training

(iii) 'Spilling down' new attitudes to supporting management levels, Juran believes that management is largely responsible for all quality problems.

Juran has formulated 10 steps to inculcate continuous quality improvements. His views are presented in figure 6.10.

Quality as a Cohesive Policy for Achieving Excellence

Figure 6.10: Juran's Ten Steps to Continuous Quality Improvements

Step (1)	Create awareness of the need and opportunity for quality improvement
Step (2)	Set goals for continuous improvement
Step (3)	Build an organisation to achieve goals by establishing a 'Quality Council': by identifying problems, selecting a project, appointing teams, and choosing 'facilitators'
Step (4)	Give everyone training
Step (5)	Carry out projects to solve problems
Step (6)	Report progress
Step (7)	Give recognition
Step (8)	Communicate results
Step (9)	Keep a record of successes
Step (10)	Incorporate annual improvements into the company's regular systems and processes and thereby maintain momentum.

Tools and Techniques
Product Design/Manufacturing Processes

In a rapidly changing turbulent world, the general axiom in regard to introducing new products would be "he who launches first wins". The firm which goes first to the market with a 'captivating new product' would certainly enjoy a significant advantage over its rival firms.

Some of the highly effective, proven tools and techniques adopted by leading American firms like Hewlett Packard, Motorola, Beckman Instruments are as furnished below:

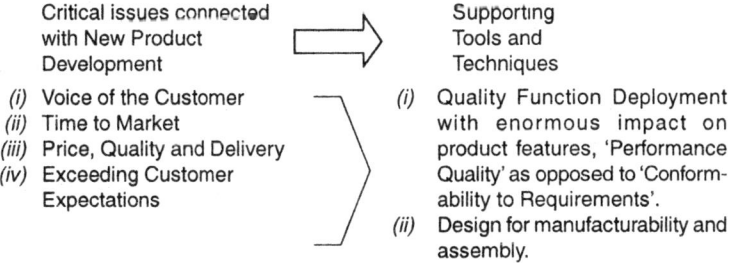

The concurrent engineering aspect of 'design for manufacturability' involves operational-staff/people heavily in the design and development of the product. Firms, which ensure involvement of manufacturing, purchasing and supplies people quite early in an intensive manner in the actual design process, would be in a position to develop superior products and put them on the market very quickly at lower costs. Thereby such firms would be in a position to offer better products which are highly fascinating and acceptable to the consumers.

Some of the winning firms like Toyota, Sony, Toshiba etc., and leading firms in the USA and Europe have acquired an inherent belief in 'people empowerment' and all the employees focus attention on four important aspects outlined hereunder:

(i) Reducing Costs
(ii) Cutting Lead Times
(iii) Improving Quality
(iv) Raising Productivity

The above mentioned firms have learned to function in a 'continuous improvement mode!' They explore every possible avenue to make spectacular improvements in order to remain in the forefront for ever.

The foremost aspects connected with 'continuous improvement' depend heavily on two factors which are similar to two sides of the same coin. These are:

(i) Implementing 'just in time' exercise (the catalyst to uncover waste)
(ii) Total Quality Control's problem-solving tools (tools to eliminate waste)

Here, a firm could receive substantial benefits only when these are implemented together. Experiments at Hewlett Packard have revealed that implementing 'just in time' exercise without 'Total Quality Control' or vice versa, would be like standing on one leg.

Just in Time Exercise

(i) Just in time automatically identifies waste and it is eliminated gradually.

(ii) The arrangement facilitates removal of waste in small increments without proving harmful to the improvement process being carried out and without risk continuous improvements take place.

Total Quality Control

Feigen Baum, when he was associated with the General Electric Company provided the following definition of TQC.

"Total Quality Control is an effective system for integrating the quality development, quality maintenance and quality improvement efforts of the various groups in an organisation so as to enable production and service at the most economical levels which allow for 'full customer satisfaction."

It incorporates a holistic set of ideas about quality which goes beyond the idea of quality 'as performance to some set of specifications.' TQM philosophy endeavours to excel on all dimensions that are important to the customer. After the 1990s, 'quality' was being perceived in terms of "total commitment from all areas, including the supply chain!"

By and large, the TQM philosophy encompasses 4 basic elements:

Customer Driven Quality

It covers external customers who purchase products/services and internal customers who get the output. It requires the "organisation" to listen to the voice of the customer.

Leadership with Greater Involvement

There must be senior management commitment to quality. Senior executives should take personal charge of managing quality.

Employee Involvement

TQM will succeed only if company-wide employee involvement and empowerment exists.

Continuous Improvement

It is the philosophy which perceives 'quality improvement' as an ongoing process of several incremental improvements.

Example: All Toyota employees in their respective functions pledge to the principles:
- *(i)* All employees must contribute to "Kaizen" (means improvement)
- *(ii)* Endeavour to improve corporate robustness so that "Toyota" will be able to flexibly meet challenges.

Focused Quality Management

Necessary steps for a proven 'four step' approach for effectively implementing focused quality management is outlined below:
1. Prepare
2. Plan
3. Deploy
4. Transition

The above approach could be suitably tailored, depending upon the uniqueness of each organisation.

Factors to be Considered
1. Process improvement is strongly connected to strategic business goals of the firm.
2. The accent is on 'improving business process', instead of attending to non-integrated projects.
3. The efforts would constitute further 'build up' on existing achievements.
4. Upper and middle managers are also inducted. They become highly committed and actively participate in 'improvement processes'.

5. The entire organisation with staffers get involved in the 'implementation programme.'

Emphasis on Focus

Strict 'focusing' in regard to quality initiatives produces spectacular results. Focus is indeed 'sharpening thinking' to a fine point consistent with organisational objectives.

Strategic goals could be effectively achieved by identifying the 'critical success factors', which are considered highly important for achieving business success. Things having no real value to achieve strategic goals will be relegated to the background.

Achieving Measurable Results

A 'Quality Leadership Team' (QLT) specifically constituted by the modern firm would reflect upon all important decisions and would also undertake process improvement implementation.

Certain important criteria for selecting process improvement projects would cover:

(i)	'Cost' of the project	→ Can it be done given the available resources?
(ii)	'Return' on investment	→ Will the benefits be worth the 'costs'?
(iii)	Visibility	→ Will it demonstrate tangible results?
(iv)	The frame	→ Can it be accomplished in a reasonable time ?.
(v)	Long Vs. short-term	→ Is there a mix of quick and long-term projects?
(vi)	Difficulties if any	→ Can it be undertaken without obstacles?
(vii)	No. of products/services affected	→ Will it enhance cross functional focus?

The 'Quality Leadership Team' may have to decide the extent to which a particular process requires improvement. The ultimate objective would be to serve the prospective customers in the best possible way, and also to attain the 'strategic objectives' the firm had already formulated.

Figure 6.11: Process Improvement Initiatives

[Figure shows three vertical axes on the left labelled High to Low: "Complexity of Change", "Resource Commitment", "Dependence on Bench-Marking". On the right, a diagonal arrow ascends through three stages: Refine, Redesign, Re engineer. Below are three horizontal arrows labelled: "Gap between Current and Future State", "Technology to Leverage Change", "Return on Investment".]

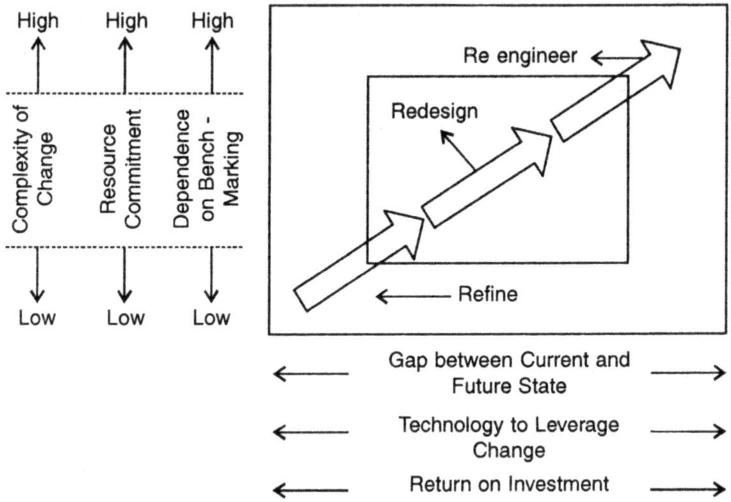

The figure 6.11 provides details relating to the three main categories of 'process improvement initiatives' a modern firm could advantageously consider for implementation.

These are:

(i) Refine — To undertake all possible incremental improvements at the outset.

(ii) Redesign — Effecting changes in the order or way in which existing activities are carried out.

'Redesigning' projects as such involves greater and more complex changes and the 'span' of improvement is larger and 'bench-marking' is involved to a great extent.

(iii) Re-engineer   By carefully assessing what has to be done, and formulating the best way to do it.

Its goals pertain to producing dramatic improvements by identifying and bench-marking the key business processes. It puts people, processes and technology on the same team and focuses them on a common goal!

Advanced Techniques
Benchmarking

'Benchmarking' is making intensive search at the outset for the 'best practices' found in any business activity. It involves undertaking systematic studies on how other leading firms tackle specific processes and also record in terms of productivity, quality and value, high levels of performance (in different departments) and achieve spectacular results. Such information would help the aspiring firm to gain valuable insights and propel them to set up challenging goals to achieve unique excellence.

The extraordinary growth of interest on 'bench-marking' has been triggered by the success stories of innumerable improvement methods adopted by the Xerox Corporation.

Rank Xerox was arranged by the copying machines produced by the Japanese at a lower price, and after necessary verification of their quality standards, began to re-organise their approach by learning from the 'best practices' in vogue.

The concept of bench-marking provides an opportunity to learn several intricacies from the experience of others:

1. It greatly facilitates development of an improved mindset among the staff as to make important contributions to competitiveness.
2. It provides greater understanding of the 'best practices' in vogue, after identifying organisations with superior performance.

3. It provides greater insight into improved processes by approaching organisations to form 'bench-marking partnerships'.
4. Besides challenging the existing practices within the business, it provides suitable guidelines for setting up goals based on actual facts.
5. Instead of depending on guess work, it gives an excellent overview of new targets to be achieved in the key areas of a firm (after observations). Therefore, bench-marking is considered to be the most sensible way of pulling up performance by the boot-straps!

Need for Capturing the Spirit of Six Sigma

During the year 1997 *six sigma* became the buzz word, and it continues to remain so. It is a complex system of statistically derived performance measurement.

The Greek word '*sigma*' here denotes attaining almost 'perfect quality' and as per the target enunciated only 3-4 defects are permitted for every million operations. Measurements give due regard to the customers perception of what is 'critical to quality' in the product or service provided by the firm.

Six sigma has been the 'key success factor' in several major businesses like Motorola, GEC, Allied Signal etc., and it can bring about organisational transformation more rapidly than any other methodology. Six-sigma mainly relies on 'excellent feedback mechanism' besides relying on factual data, and statistical measurement techniques so as to ensure high quality decision-making.

As General Electric Company calls it, six sigma is "the road map to customer impact" and a way of "striving for perfection" in meeting 'customer expectations'.

The basic assumption is that the customer needs consistency on the quality variables he (or she) is sensitive to, and six sigma requires production (or other value delivery) processes to be

Quality as a Cohesive Policy for Achieving Excellence

controlled so tightly as to reduce output variation to a level that results in no more than 3-4 defects per million opportunities.

Six sigma is essentially probability based. Here an opportunity is defined as a chance of failure in specified limits for the variable (for example a toilet soap's total fatty matter content) the probability of which is assumed to be normally distributed around the mean.

This is presented in the 'Bell Curve Chart' (with the variables value on the X-axis) as figure 6.12.

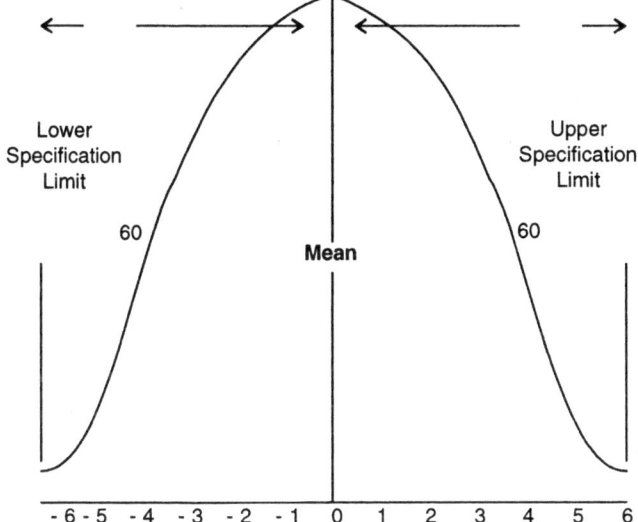

Figure 6.12: Six-Sigma Quality: The Bell Curve

One could see that the probability peaks at the Mean, and there is likelihood of missing the Mean once the variable deviates from this value.

To adopt six sigma, the manager/technologist should arrange to 'rejig' his processes to squeeze the 'bell' within a 'tight frame' so that the probability of staying within 'Customer Pleasing Limits' is as close to the ideal certainty of: 1 as possible: 0.999997.

Such a pursuit of statistical perfection confers innumerable benefits on the aspiring firm.

(Source: "Diagram" presented in *Business Today* Magazine dated 12.10.2003).

The Six Sigma Programme

The *six sigma programme* has been built around a 'problem solving methodology' called "M.A.I.C." which stand for:

(i) Measure
(ii) Analyse
(iii) Improve
(iv) Control

Barbara Wheat, Mills and Carnell explain the strategy in the following lines:

"The methodology is used on 'chronic problems' selected for 'black belts' to use on." These black belts are people who have gone through a 'training process' and completed projects to gain certification in the six sigma problem solving methodology. The 'projects' are selected by "champions" (who are selected by the leadership team) to address 'chronic problems' in strategic alignment with the company's business objectives.

In consonance with Juran's line of approach, 'six sigma' could be called a breakthrough strategy, and any firm adopting it is engaging in a change programme. Besides, some 'black belts' are selected to undergo additional training, and they are certified as 'master black belts'. Master black belts, besides being the mentors of several black belts, also create new black belts.

There are also some workers known as 'green belts' and trained in 'six sigma skills' often to the same level as black belts. The green belts help to bring the new concepts of six sigma right to the day-to-day activities of the business.

As a group the champion master black belts, black belts and green belts tackle several chronic problems in the form of projects

Quality as a Cohesive Policy for Achieving Excellence

and all the projects are strategically aligned with the objectives formulated by the firm.

The rank and file of the organisation is well-versed with deficiencies confronted, and they make every possible endeavour to tone up the healthy growth of the organisation.

Basic Equation on Six Sigma

The concept revolves around a basic problem solving equation

viz., $Y = (f) X$

or $Y = (f) X_1 + X_2 + X_3$

This equation explains the relationship which exists between a dependent variable – Y and independent variables: the Xs.

Y is the output, which relates to the 'final product'.

The output is actually a function of different inputs (the xs). From this it becomes obvious that only by controlling all the inputs one could completely control the output. Ultimately the 'firm' is able to arrive at an optimised solution.

Jack Welch as Chairman of General Electric Company achieved remarkable success by disseminating certain key initiatives throughout his organisation. These are:

(i) With 'six sigma', he imposed mandatory training in the technique for GE managers and he linked rewards to reaching proficiency.

(ii) Introducing the change acceleration programme

(iii) Arranging cross fertilising of all 'best practices' on a global scale and making several practices with boundarylessness. (infinite scope for development).

(iv) Every division of GE would be reinventing itself in e-commerce terms, before a competitor overtook it.

(v) Work out and six sigma had been folded into the digital processes, and thereby produced a new IT-consultancy business for General Electric, in which several leading customers and providers followed the work out discipline involving brain storming sessions.

Besides, Jack Welch had the inner conviction that true business would flourish only by maximizing the 'intellect of the organisation". He created a marvellous learning organisation where high performing managers were real embodiments of corporate values. He had great sagacity in capturing the new trends, and capability to re-model his business accordingly. Besides, he had no hesitation in weeding out 10% of staffers who were not making worthwhile contributions to the healthy growth of the firm.

His famous dictum was: General Electric had to be no. 1 or no. 2 in the market, and, if not, measures to 'fix, close, or sell' would automatically follow.

Training Programmes for Quality

A well-planned and systematic training programme is the most important prerequisite for any aspiring organisation keen on achieving high standards of product excellence.

The training programmes should be suitably organised keeping in view:

(i) The constant changes in technology.
(ii) The structural and environmental factors concerning the organisation.
(iii) The different categories of people who require adequate exposure to the latest trends and techniques.

Before formulating the actual 'training objectives', the modern firm should undertake an in-depth study of certain factors:

(i) How the feedback mechanism' to ascertain the customer requirements, latest trends in fashions/fads etc., is functioning.
(ii) Identifying the key areas which require improved performance/to impart training and enhance efficiency of employees.

(iii) The need for training and acquainting employees with new skills for new technology, higher operations and efficient performance/training by objectives.

(iv) The kind of new procedures to be formulated, and necessary provisions to be made.

(v) Improving 'motivational climate' and reinforcing 'participative culture'.

(vi) 'Training' should be 'need based' with separate training programmes for:

   (a) Augmenting core skills

   (b) On-the-job training

   (c) Suitable training programmes for acquiring advanced skills/multi-skills.

The training programmes should be suitably tailored to achieve certain ends:

(i) High morale and commitment among all employees.

(ii) Attainment of necessary competence and job satisfaction.

(iii) Excellent participation in team work and a high sense of team spirit.

Ultimately, the results of the training should bring about

(i) Greater productivity.

(ii) Cost reduction + high profit margins.

(iii) Projecting a better organisational image.

(iv) High quality of work life and the firm acquiring adequate competitive edge over other firms.

To assess the effectiveness of the various training programmes, evaluation should also be undertaken periodically by the aspiring firm.

7

Harnessing Managerial Dynamism: Vision for 21st Century

If 90% of the "Human Energy" is out there trying to resuscitate, revive and revitalize-the Low System of Management, that would not surprise me. But if we have 10% available to work on creating something 'really new', then that may be more than adequate.

—Dr. Peter M. Senge-MIT

UNCERTAINTY IN GLOBAL BUSINESS

Every CEO, or top executive is aware that he (or she) has to operate in a global environment of increasing chaos. As such the CEO is confronted with two sources of change:

(i) The traditional change initiated by the firm.

(ii) The external change over which one cannot have control.

Besides, risk factors have considerably increased on account of growing inter-connectedness among industries functioning in different parts/countries across the world.

While functioning in an era of uncertainty, creating a stable organisation would require much sagacity and specialised skills. The employees who grow fearful of 'down spring', in general, function amidst tension and insecurity.

The CEO at the top should possess sophisticated emotional skills in order to instil the required support and confidence among the staffers.

In the 21st century milieu, several functions such as planning, forecasting, staffing, budgeting etc., cannot be pursued as per strategic plans in a modern firm, since it is rather difficult to bring future into focus. Therefore, it becomes necessary to prepare for the future with new modes of production. Besides, there arises the overwhelming need to strengthen the quality of relationships among people while confronted with negative dynamics. People should develop community spirit, and should be engaged in meaningful work, which should transcend individual concerns, and develop new capacities and undergo suitable drills. They should be capable of tackling new scenarios while a sense of uncertainty prevails.

Need for Individual Attention

The CEO and the top executives should establish personal contact with staffers. They should regularly hold conversations with people, incharge of key areas, experienced workers, innovative people, consultants, and so on. When the staffers of a modern firm feel cared for, they are willing to contribute more towards the healthy growth of the firm. When quality relationships are strengthened, 'better knowledge-exchanges' take place!

In case the firm has no alternative, except shrinking the workforce, it should be undertaken only with considerable flexibility without sacrificing dedicated, trustworthy loyal and smart people!

International managers struggle much in finding suitable solutions. They cannot purely depend upon technical rationality alone. They have to activate more their intuitive brain power for making action-learning approaches.

As such, managing 'uncertainty' poses several challenges. 'Uncertainty' may mean:
- Not precisely determined, indefinite or unknown!
- Putting one in a state of indecision.
- Ambiguous, undependable.

The modern managers are lost in a haze, since the events taking place right now are not precisely determined, and are ambiguous, contrary to the past when developments were neither uncontrollable nor unmanageable. When they have to deal with increasing complexities of the business world, they are vaguely aware of their responsibilities and quite often lack confidence, since the nature and the magnitude of the problems to be tackled as such, are not clearly known, and not even the expectations of their superiors.

The modern managers must be mentally and emotionally well-equipped in order to steer clear of trouble shooters, in the turbulent business milieu.

Ralph Stacey in his remarkable book: *'Managing Chaos'* (1992) has stated:

"The new approach means accepting that you really have no idea about what the long-term future holds for your organisation. Further, the new approach is about positively using instability and crisis to generate 'new perspectives' provoking continual questioning, and organisational learning, through which unknowable futures can be created and discovered."

Besides, he states "the structures and behaviours necessary for 'stable normal management' have to 'coexist with the informality and instability of 'extraordinarily management' which is necessary to cope with the unknowable.

In this regard, Stacey further indicates that "there is too much emphasis in management today on stability, regularity, predictability and cohesion". Instead, he believes that business should lay accent on the "management of bounded instability" in which behaviour and systems go completely out of control and, sometimes, may also bring about self-destructive reactions. In the new management paradigm, chaos is treated as a natural part of change and should also be effectively managed.

Need for a New Mindset

George Bernard Shaw, a towering intellectual figure made a startling-observation:

"All progress indeed, depends on unreasonable men!, since only unreasonable men endeavour to 'adapt the world to themselves', instead of adapting themselves as any rational human being."

On account of globalisation the world is becoming highly complex and less predictable, and as a result the modern organisations have also become highly complex, with several challenging issues. The greatest challenge before modern managers would, therefore, be the 'capability' to confront them in a highly dynamic international arena, and intuitive skills, which they possess in abundant measure, to deal with complexity, change and ambiguity.

The personal mindset of the modern manager would reap rich dividends, only if he could utilise the activities of his right brain responsible for the intuitive skills, so essential for the current scenario.

Ultimately, it could be observed that the following basic capacities would actually constitute the essential ingredients of a 'new mind-set':

(i) Tackling problems in a comfortable manner even with ambiguity

(ii) Inherent capacity to utilise any uncertainty as an opportunity for further growth.

(iii) The innate tendency to translate every opportunity to look for new product lines, improvements upon existing products, (eg.: the first cigarette with a filter-tip), services, strategies, structures, etc.

(iv) Ability to perceive potential obsolescence of different products services, strategies and structures.

(v) Curiosity to view things diffcrently and think about how the future scenario may unfold.

It is an acknowledged fact that the higher one goes up in the hierarchy of the organisation, the more complex become the decisions, and the data essential to draw worthwhile conclusions become meagre. Here–

(i) Capability to make correct judgements and

(ii) Ability to obtain a clear vision of the unfolding scenario become the most crucial factors.

With sparse information and high-risk factors, middle managers invariably make it a habit to pass on the decision-making to the higher level.

In such cases, mostly tough decisions are taken by top level executives and the 'CEO'. The CEO and 'top executives' should make every possible endeavour to contemplate an organisation that would enhance the health of overall business in a climate of rapid change.

Though the organisation should be flexible, and responsive enough in this new challenging milieu to survive and grow, it should not lose sight of the need to conform to certain 'strict tactical requirements' essential for 'short-term goals'.

While operating in an international environment, they have to formulate the most effective 'strategic-outfit' taking into account the following factors:

(i) In the 21st century, several successful enterprises tend to utilise the technologies at the speed of light so as to increase efficiency combined with time management. Here it becomes utmost necessary to explore beforehand, suitable markets in which any modern firm could actively participate.

(ii) To become a 'high performance firm', there is need to formulate a strategy which would ensure '*continuous improvements*' in respect of product features and service details.

Success could be well within reach by inducting distinctive capabilities in the organisational set up'. Besides the CEO, top executives and staffers must have a 'growth intoxicated' mindset. Hence, the old adage: 'slow and steady wins the race' will no longer be found applicable!

As such, people all over the globe have acquired a tendency for quantum jumps in their standards of living. Besides, the market economy has started imposing high consumption life styles with greater expectations of 'adduced value to products'. In view of these developments, 'a totally new managerial mindset' will have to be mastered.

In order to grow rapidly on healthy lines, the following factors should be accorded priority by the modern firms.

(i) It should well assimilate its customer base and customer needs for different product categories.

(ii) Acquiring vital information of the specific industry-group with details of purchasing decisions in different cultures/countries, advertising techniques, most essential demand factors, etc.

(iii) Since the modern firm may have to encounter consistently several competitors in the same product lines from different countries, the regulatory environment of each country (rules/regulations) should be thoroughly assimilated at the outset.

Since there are approximately 180 to 200 global territories/countries all over the world, the global firm should select the core market places suitable from the point of view of achieving greater success.

In the emerging global economy, a firm's competitive survival ultimately depends on how well its 'strategic and managerial techniques' could confront several rival firms from different parts of the globe.

(iv) Lead markets are situated in countries having capabilities to initiate specific innovations or improvements which would subsequently cause 'ripples across the global economy.'

(v) The CEO, in consultation with the top executives of the firm, could suitably plan product launches after identifying suitable market places on the basis of signals received from the lead markets of 'advanced innovative countries".

(vi) The CEO of a global firm should incorporate adequate flexibility and adaptability in terms of resources for the 'global build-up'. Besides, he should apportion all the resources at the firm's disposal in the most judicious manner by adopting a 'superior focus on key markets', selected at the global level.

(vii) The global manager/CEO should learn the 'art of change' in an uncertain world. In this regard, Andy Grove, renowned as an expert on 'Change', has aptly observed; "You need to try to do the 'impossible' to anticipate the unexpected! And when the unexpected happens, you should 'double your efforts' to make 'order' from the disorder it creates in your life. The motto I am advocating is 'Let chaos reign,... then rein in chaos".

It could be called the 'chaos-control theory of organisational change', which actually involves the 'EBB' and flow of letting go, and also 'taking control'. It may also be necessary to destroy in toto the old-culture in order to embed 'new values, norms, behaviour' in an organisation's corporate culture.

In this regard, the CEO has to depend more on his 'left brain process', and his actions should be highly rational and analytical.

Executive Excellence through Organisational Learning

It is utmost necessary to accelerate 'Organisational Learning' on the whole, besides establishing a managerial talent-pool in the global firm.

To accelerate the learning process Jack Welch, the renowned C and MD of GEC, has recommended certain measures.

(i) A group of 30 to 60 employees would be taken to a conference-site. The co-ordinating boss would present an agenda to carryout a brain storming session mainly to glean bright and concrete ideas on quality, product features, productivity and so on for ensuring 'accelerated improvement'.

The aspiring CEO (or entrepreneur/manager) should suitably arrange to interview a few select firms which have achieved spectacular success in the global front with a view to unravel certain distinct factors and characteristics, which are mostly attributed to their success.

It is gathered that the General Electrical Company, USA always generates a flow chart outlining every step involved in bringing out modified (improved) product lines. Indeed, such 'process mapping' would help a great deal a modern firm in 'revitalising its multifarious product lines' and also facilitate its growth on healthy lines.

(ii) Only by means of 'gross pollination of ideas', efficient plants and firms could be transformed into international production centres. Besides, there is the imperative need to focus greater attention on research and development efforts so as to generate useful innovative ideas which would lend excellence in a consistent manner.

The aspiring global firm should also make concerted endeavours to undertake periodical evaluations of its standards against the 'best breed of competitors', in the product lines in which it is specialising. Due regard should also be given for 'changing customer expectations'.

(iii) Prof. John P Kotter of Harvard University has wonderfully illuminated the daunting complexity of the CEO/or General Managers' job.

The principal tasks of leadership identified by him are:

(a) Establishing 'direction', developing 'a vision' and 'strategies' for the future of the business.

(b) 'Aligning people' – getting others to understand, accept, and line up in the chosen direction.

(c) 'Motivating and inspiring people' while managing processes are about 'control and monitoring'; Leadership processes accomplish their energising effect not by pushing people alone in the right direction but by satisfying very basic human needs for achievement, belonging, recognition, self-esteem, a sense of control over one's life and living up to one's ideals.

He has further defined management roles as:

(a) Planning and budgeting 'short to medium-term targets'

(b) Establishing steps to reach them and allocating resources

(c) Organising and staffing, establishing an organisational structure, to accomplish the plan, staffing the jobs, communicating the plan, delegating responsibility, and establishing systems to monitor implementation.

(d) Controlling and problem-solving, monitoring of results, identifying problems, and organising to solve them.

(iv) Peter Senge at MIT's-'Sloan School of Management', has intensely studied, how an organisation "develop adaptive capabilities", amidst increasing complexity and rapid change. Senge has laid stress on the need behind obtaining competitive advantage (by every modern firm) from "continuous learning" – both individual and collective.

Since the technology of the "information age" is radically changing, firms should radically restructure their operations, besides making constant improvements in their learning process.

(v) Prof. Warren Bennis, a leading management thinker has provided thought-provoking viewpoints on leadership qualities.

(a) Successful leaders follow almost an universal principle of management as true for "orchestra conductors, army generals, football coaches, and school-superintendents, as for corporate executives!"

(b) What managers expect of their subordinates and, "the way they treat them" largely determine their performance and career progress.

(c) At the same time, leaders are realistic about expectations. Their motto is: "Stretch!..... don't "strain"! Pretend you are training for the Olympics, where 'easy' does it. If you pull a muscle in today's game, you sit on the bench for tomorrow.

(d) Leaders possess the noble factor-"optimism, faith and hope". Ronald Reagan is a good example of this boundless optimism. It is stated in a Chinese proverb: "That birds of 'worry and care' fly above

your head, this you cannot change; but that they build nests in your hair, this you can "prevent".
   (e) Invariably great leaders formulate 'long range plans' to 'see the long-term view'. Long-range plans, set up for periods varying from 10 years to as much as 250 years, are in vogue in different parts of the world, and the Japanese firms are renowned for their 100 to 250 year-long planning programmes.
   (f) Several leaders endeavour to "balance the competing claims of all the groups who have stake in the corporation" (groups encompass consumer forums, constraints of government policies, trade unions, etc.). Decisions are taken only after studying the impossible impact on public opinion. Several leading corporations opt for "stake holder symmetry" as the foremost principle to be followed amidst the complexity of the 21st century milieu.
(vi) Several management experts have observed that "learning is a life long process". In any organisation developing the skills and abilities of the staffers is of paramount importance to achieve phenomenal success. In order to sustain the competitive advantage, the modern firm should substantially invest on its people.

It has been estimated that, all over the globe, over the next 10 years opportunities may open up for 1.3 million new high tech jobs, but there will not be sufficient skilled workers to fill them. In this context, "employee-learning and development" is of paramount importance to achieve substantial results.

Once the top management has formulated the 'vision', true commitment begins, and the staffers in coordination will have to take necessary action. The 'vision' itself may expand over a period of time, though its 'core values/ideologies may' remain the same.

However, the organisation as a whole should courageously tackle unforeseen events and obstacles presented during the course of implementation, and move full steam ahead encompassing a much broader community.

In any dynamic organisational set-up, visioning becomes an ongoing process. When the staffers share a 'vision' and firmly believe in its purpose, invariably it generates tremendous excitement, energy and passion for achievement. As a result, "the organisation is able to build a strong reputation for products of excellent performance standards and expeditious services." This would automatically enhance the number of loyal customers.

Importance for Talented Personnel

The aspiring global firm should acquire greater fascination for inducting talented personnel in its organisational set-up. The CEO should develop adequate obsession to reach the top by becoming a genuine connoisseur of talent. Tom Peters has categorically stated that the modern CEO should not settle for anything less than the best. Though top talent in any field may prove highly expensive 'mediocrities' should not be allowed to fill the chairs in a growing organisation. Indeed, a talent obsessed enterprise in the 21st century milieu would certainly reap rich dividends, since the talented people who are well-nurtured by the firm would contribute quite substantially towards its healthy growth.

Risk of Downsizing

The modern CEO of a firm may be confronted with the problem of down sizing (assets or employees) on account of *macro economic factors* and when the economy is facing recession. Surveys conducted by the 'American Management Association' have revealed that "downsizing" has not produced tangible results.

The disadvantages on account of reduction of human assets are greater in the long run for an aspiring firm though temporarily (in the short run) the margin of profits may indicate a minor upward trend!.

Major Disadvantages of 'Downsizing'

1. It leaves the organisation in a 'shell shocked position', which may have a demolishing effect.
2. Loss of company's "innovative ability" and "absence of improvement in product quality."

In view of these factors, the CEO may consider 're-allocation of jobs' instead of resorting to downsizing measures. This would certainly facilitate the firm to create new products required by customers periodically in addition to improving the overall tempo of activity within the organisational set-up.

Need for 'Double Loop' or 'Open Loop' Learning

In corporate governance the CEO and top executives should always endeavour to drive the business forward, while keeping the firm under their prudent control.

Besides being sensitive to short-term pressures and local problems, they should not lose sight of the following factors:

(i) They should constantly keep in mind the long-term objective of the healthy growth of the enterprise as a whole with sufficient knowledge of the internal functioning of the firm.

(ii) Broad trends at the national and international level should receive greater focus with sufficient information on 'emerging competitive trends.'

(iii) It should act responsibly towards its employees by ensuring
   (a) Job enrichment
   (b) Team spirit

      (c) Implementing effective reward and recognition systems
- (iv) Formulating a clear 'customer service strategy' by utilising interpersonal skills.
- (v) Exceeding 'customer expectations' (customer delight), and going the 'extra mile' to leave memorable impressions on consumers.
- (v) Create a reputation in the market place for exceptional service and thereby ensure building 'long term viability' for the business.

These characteristics could be inducted in an aspiring firm if Bob Garret's 'double loop learning' is undertaken.

The following figure 7.1 provides details on the functioning of 'double loop learning'.

Figure 7.1: Double Loop Learning

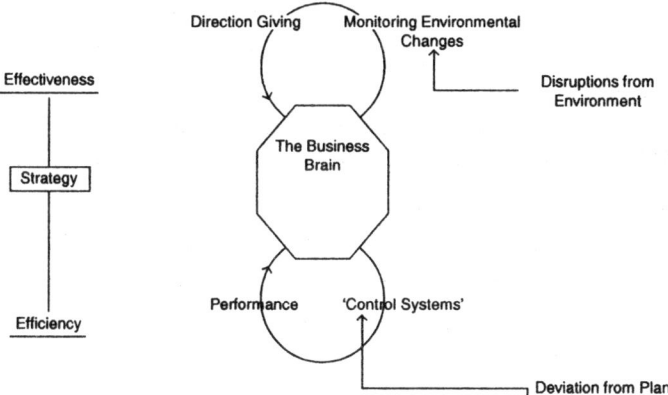

In this context, it is worth mentioning Professor Reg Reven's famous maxim which lays accent on 'providing focus on learning', that would provide opportunities to all the departments of an aspiring enterprise to 'simultaneously learn' and also 'survive'.

The following formula is an excellent maxim to be followed:

$$L \geq C \text{ For Survival}$$

Here, L = Learning
C = Rate of change

Prof. Reg Raven has elegantly summed up the indispensability of learning as: "Learning must be greater than or equal to the rate of change".

Characteristics of Highly Successful Companies

Professor Amin Rajan of City University Business School provides excellent perspectives on the distinction between "management" and "leadership".

He indicated:

(i) Management is about 'now'; leadership is about the future.
(ii) One implements 'goals', the other 'sets' them.
(iii) One relies on 'control', the other inspires 'trust'.
(iv) One deals in 'rational processes', the other in 'emotional horizons'.

Prof. Rajan, after intensive research of several firms which have achieved dominance in the international arena, concludes that they possess the following five key ingredients:

(i) Their leaders are visionary and enthusiastic champions of 'change', who have communicated their 'business goals' throughout the company and generated the necessary commitment at all levels.
(ii) They have flexible motivated employees capable of performing multiple tasks at different levels of responsibility.
(iii) They learn from the customers, competitors, suppliers and academicians.
(iv) They have a culture of "innovation" that seeks continuous improvement in products and working methods.

(v) Their business processes and practices seek to achieve 'low cost product customisation', while retaining a clear edge in price, quality and service. They have inverted traditional manufacturing logic by securing (opting) for "batch production" with all those advantages commonly associated with "mass production" and "standardisation" Collins and Porras in their in-depth study of some leading firms have come across certain special attributes peculiar to these in regard to the following factors: (Table 7.1).

Table 7.1: Blending of Certain Characteristics

While tackling 'critical issues' most of the leading firms blend certain characteristics.

	On the One Hand	Yet-On the Other Hand
(a)	Purpose beyond 'profit'	Pragmatic pursuit of profits
(b)	A relatively fixed core ideology	Vigorous change and move
(c)	Clear vision and sense of direction	Opportunistic grouping and experimentation
(d)	Ideological control	Operational autonomy
(e)	Investment for the 'long-term'	Consider 'demands' for short-term performance
(f)	Philosophical visionary, futuristic approach	Superb daily execution 'nuts' and 'bolts'

Policy Deployment

As per guidelines provided by Edward Deming, the Japanese had formulated the system of "policy deployment" or "management by policy". The foremost logic in this approach is one of 'cascade consultation, involvement, and consensus on policy and strategy".

The exposition in regard to 'Policy Deployment Processes' provided by Peter Vickens, management expert is worth examining.

The following figure 7.2 presents the policy deployment process of a modern firm as portrayed by Peter Vickens.

His pragmatic approach indicates that objectives are integrated with business strategy, and permeate throughout not

Figure 7.2: Policy Deployment Process

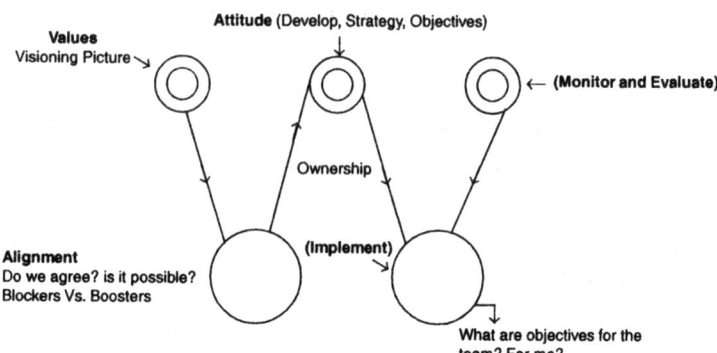

only 'top down' but 'upwards' and laterally, and this can only be done by involving people at all levels. The very process is part of their development.

'Top management' as a team needs to begin the process by determining the corporate objectives; but in cascading through the organisation, the process has to be not one of 'top down' imposition, but of asking people 'what they can do' to contribute, discussing it, assessing the impact on their colleagues, and agreeing not only on the objectives, but also the means of achieving them; and the measurements to be applied.

At all stages discussions between those affected take place. The Japanese use the term: *"Koshin Kanri"* (Policy Deployment) for this process, and the discussion and debate over the objectives is termed 'Catch Ball'. It is a superb process for determining what really are the 'key objectives', and for ensuring that they are genuinely shared.

Helix of Never-ending Improvement

Managers should have clear ideas relating to:

 (i) What are the employees required to do?
 (ii) What are standards of performance being expected?

Only when there is greater awareness, and an unimpeded understanding of these ideologies among all staffers, the CEO,

Harnessing Managerial Dynamism Vision for 21ˢᵗ Century

top executives as well as managers, the firm can hope to get what is expected of them.

Needless to say that, an inspiring vision formulated could be achieved only by active employee participation. This again depends to a large extent on the education and training received by the employees in consonance with the needs and expectations of the firm.

Accordingly, all the employees of the modern firm should be trained to:

- (i) E: Evaluate → Assess the situation and define their objectives
- (ii) P: Plan → Achieve all objectives
- (iii) D: Do → Implement the plans
- (iv) C: Check → Ensure that the objectives are uniformly achieved
- (v) A: Amend → Take suitable corrective action when required.

Training imparted on the basis of these concepts would help to implement disciplined management within the organisational framework.

Here the staffers at all levels must perform sincerely. "What they say they will do". It would also mean in whatever they do they would systematically go through the full process of: "Evaluate, Plan, Do, Check and Amend" rather than the traditional approach of immediately beginning to act instead of evaluating at the outset.

This line of approach would certainly help the creation of a *"Never ending-Improvement-Helix"*.

This 'Helix' is presented in figure 7.3.

Such a methodology of approach should be backed up with:

- (i) Excellent project management
- (ii) Appropriate planning techniques and
- (iii) Correct problem-solving methods

Figure 7.3: Never Ending Improvement Helix

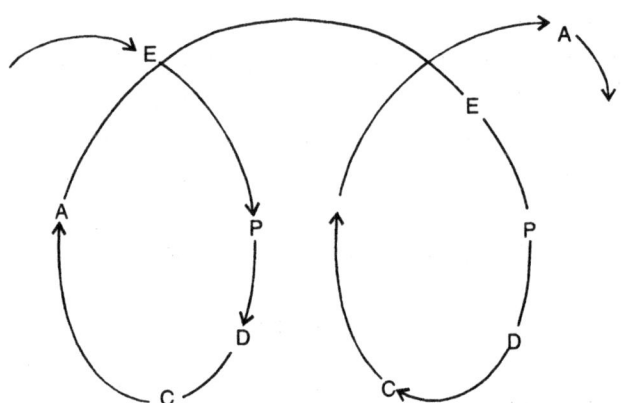

The employees become well-versed in such scientific concepts through suitable training programmes of short duration.

These 'continual improvement techniques' when combined with effective leadership would facilitate superior performance, resulting in organisational excellence and greater market share.

When every individual in an organisation clearly knows what he or she is supposed to do, the long range *'organisational objectives'* as well as the tactical short-term objectives could be well-achieved.

Special Competencies Required for Leadership

In the 21st century environment leaders can not afford to spend 70% of their time on getting the job done as was customary of traditional leadership.

Leaders of modern firms are required to spend more time on 'innovation' and 'productivity' related tasks.

Leaders have to develop competencies such as better perception, emotional maturity and empowerment.

Since innovation has become the 'new frontier' for leaders, they should work through people, and become much more 'people centric'. In addition to being efficient and effective, they should shift their competency focus to creative thinking and endeavour to release the creative spirit of people in teams. In this regard, it is essential for a leader to create a culture where people feel comfortable offering ideas for improvement. Besides, the leader should endeavour his best to create a cultural readiness for 'change' among all employees. By striving to create a culture that is ready for 'change' with 'willingness to thrive', several changes of decisive importance could be implemented with the enthusiastic support of all employees.

Jack Welch, CEO of GEC has aptly observed that a true leader, besides having infectious enthusiasm for infusing quality, should not constantly remain in one job for long! He should "rewrite his agenda" from time to time, in order to explore new challenges that offer intellectual growth. Besides, he has stated that a CEO should not fall into "false scenario trap" by just assuming, that things will get better! It is utmost essential to face truth in a detached manner and to thoroughly understand the realities of the situation in order to gain a clear perspective.

Swot Analysis

A study of the "best in class" companies such as John Deere, Royal Dutch/Shell, Siemens, Sterling Chemicals, Xerox, Silicon Graphics has revealed that among several techniques and approaches employed for healthy growth, "SWOT" – triggered business plans have also been given much focus and attention.

Here it is thoroughly verified that the action plans formulated by the organisation are directed towards outcome and performance measures that are in line with 'strategic objectives'.

As such SWOT analysis is quite instrumental in optimising the performance of a modern organisation. SWOT stands for 'Strengths, Weaknesses, Opportunities, Threats'.

Here, S and O → stand for 'positive attributes'

W and T → stand for 'negative attributes'

Besides, there are *two important dimensions.*

(a) Conditions Pertaining to Internal Environment
It refers to conditions which pertain to the firm's internal strengths and weaknesses. These characteristics already exist and a careful analysis would facilitate 'decision making'.

(b) Conditions Pertaining to External Environment
Here, the external features of the organisation's 'opportunities' and 'threats', when carefully analysed, provide useful ideas to tackle future problems arising out of changes in 'technology, demography and government's policy.

The following table 7.2 illustrates the 'primary internal and external conditions' – which could be advantageously studied by the 'global firm' (by using 'Swot Analysis') to draw correct conclusions.

Table 7.2: SWOT Analysis

S.No.	Internal Strengths and Weaknesses	External Opportunities and Threats
1	Horizontal Systems/Processes	Trends in Customer Value
2	Organisational Structure	Economic Trends
3	Management Capability-Parameters	Technological Factors
4	Financial Position	Rules and Regulations
5	Marketing Strategy	Competitive Situation
6	Human Resources	
7	Corporate Culture	
8	Research and Development	

Documenting "SWOT" Findings

While undertaking SWOT Analysis, one should not neglect glaring weaknesses within the organisational framework. Besides, the strengths should not be overlooked by concentrating on existing weaknesses alone.

SWOT 'brainstorming sessions' with, people renowned for 'original thinking' would prove highly helpful. While compiling the SWOT summary, it is essential that all 'strengths' are truly identified, and could be utilised to 'benefit customers'. After listing weaknesses, it is necessary to identify all the 'key areas' where 'performance could be improved'. 'Opportunities' for further growth and threats likely to be posed should be well analysed.

The following figure 7.4 shows how SWOT summary has to be used for the 'strategic plan'.

Figure 7.4: SWOT Summary

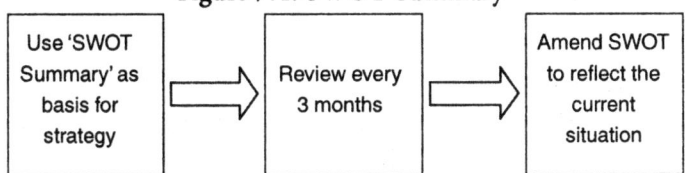

Besides helping to get a clear 'direction', the firm would be able to determine where the 'competitive advantage' lies. This 'analysis' would be much helpful in reviewing and updating 'decisions' on a regular basis.

Decision making and Sophisticated Knowledge Base

The CEO/general manager of a global firm, should develop a refined social architecture that would also be instrumental in generating 'intellectual capital'.

The aspiring firm requires inputs from a broad range of people and should therefore develop essential knowledge pertaining to:

(i) Competitive environment
(ii) Customers' competitors
(iii) Suppliers

(iv) Strategic alliances
(v) Future opportunities for growth
(vi) Future threat

By developing such a sophisticated knowledge base, the firm would be in a position to derive 'adequate feedback' which would in turn serve as a reference for evaluating new/future information.

This would ultimately facilitate the global manager/CEO to formulate 'timely business decisions'.

The overall objectives of inducting 'business intelligence process' in the global firm would pertain to certain advantages:

(i) Learning is triggered to generate adaptive behaviour
(ii) Monitoring more about external competitive environment
(iii) Injection of new information and making it more transparent and thereby facilitating organisational changes suitable for local fitness and global fitness.

As a result, the 'business information process' becomes part of the 'perpetual strategy process'.

The business information process and the underlying knowledge base would present largely a game of perspective though not with much precision.

By gaining the required 'perspective' the CEO/global manager would be able to make excellent strategic and tactical decisions.

Inducting Distinctive Capabilities

The global manager keen on achieving spectacular success, while navigating his company's international growth, should no doubt induct certain distinctive capabilities as enumerated hereunder:

1. Developing a social architecture by encouraging talented executives/brilliant employees.
2. To build reputation over time and to project a finer

image, a suitable mechanism of advertising and warranties would be essential.
3. Capability to offer high quality products at lower prices by resorting to innovation and upgradation in order to stay ahead in competitive race.
4. To have a 'learning mindset' which should be fast with richness and diversity.
5. Since speed and accuracy have become the most dynamic factors, all employees should have 'growth intoxicated mindsets' within the firm.
6. Positive reinforcement to cope with 'high consumption life models', 'retention of customers' and achieving a 'motivating climate'.
7. Greater focus on special (specific) global markets.
8. The mega trends of western influence in the Asian way (Culture) should be well-established.
9. The 21st century organisation will have to be re-invented constantly; continuous improvement of every product, process and service would be essential.
10. Several leading economies of the world have now become 'market controlled' and not 'government controlled'.
11. It is absolutely essential to pursue a policy that favours the healthy growth of 'services sector' with much diversification.
12. An innovative and entrepreneurial culture that would also recognise and nurture women entrepreneurs would be found essential.

Mergers and Acquisitions

The term "Mergers and Acquisitions" encompasses several activities such as: joint ventures, licensing, spin-offs, equity carve-outs, restructuring, alliances and so on. The crucial role of M and A activities is to facilitate firms to adjust more effectively to new challenges and opportunities. M and A would prove

instrumental in gaining access to new markets and play a significant role in 'corporate strategy'. The resultant 'synergy' would certainly confer substantial and distinctive capabilities on enterprises as outlined below:
- *(i)* Substantial cost reductions could take place on account of "economies of scale".
- *(ii)* Confers increased monopoly power.
- *(iii)* Facilitates manufacture of broader range of products and useful services.
- *(iv)* Managerial empires emerge and market networks get strengthened.
- *(v)* Gaining greater access to improved technologies, besides sharing of risks and costs associated product innovations. Such product innovations could be enhanced with 'lesser time spans'.

Intangible Assets

Leading companies inherently possess unique knowledge, which have excellent potential to provide competitive advantage over other firms. The knowledge could be found in the form of:
- *(i)* Technological development
- *(ii)* Proficiency in a specific core business
- *(iii)* Information/experience pertaining to the industrial structure.
- *(iv)* Peculiarities of customer behaviour.

Such assets may possess abundant value even though they are identified in a 'highly illiquid form', and remain imbedded in the operations of the company.

In the global context, with vast expansion in the volume of knowledge available, the problem pertains to 'sensible knowledge management' by aspiring firm and the conversion of knowledge to 'value creating forms'.

The greatest challenge is to effectively use the "knowledge based intangibles", while gaining access to global markets, and in this regard every endeavour should be made by the aspiring firm towards maintaining proprietary control over the knowledge accumulated so far. This would constitute the vital component of the strategy for having sustained competitive advantage over the rival firms.

'Intellectual Power' could be very well-termed as the new route to competitive advantage. Smart decisions emanating from collective brain power boost up success in the corporate world.

Prof. Peter Drucker has aptly summed up: "The world is becoming not labour-intensive, not materials-intensive, not energy-intensive but 'knowledge-intensive"!

Transferring Intangible Assets Into Intangible Capital

The following figure 7.5 presents how the embedded intangible assets get converted into deployable intangible capital:

Figure 7.5: Conversion into Intangible Capital

S.No.	'Embedded' Intangible Assets	Deployable 'Intangible Capital'
(A)	KNOWLEDGE Internal knowhow; specific information/excellence	INTELLECTUAL PROPERTY Knowledge transferred through Research and Development with legal protection. Property earns direct cash returns.
(B)	PEOPLE Global pool of talented individuals	TALENT Highly talented people become world class performers. They are selected and are allowed to create/exploit winning global propositions.
(C)	RELATIONSHIPS Established advantageous link between producers, suppliers and customers	NETWORKS Privileged ownership of infra-structure providing value to connected parties through direct economic benefits covering a wide spectrum.
(D)	REPUTATION 'Superior Value' proposition in delivering goods and services	BRAND Concerted endeavours to promote distinctive reputation which lowers interaction costs with customers and help to recover higher prices.

Source: 'RACE FOR THE WORLD-1999' By Lowell Bryan, Fraser, Oppenheim and Rale

Among the leading firms of the world, there is indeed growing realisation on the need for recruiting, retaining and nurturing talented people in order to ensure competitive edge over others.

However, turning knowledge and intellectual capital into reality poses substantial challenges. Knowledge management programmes organised by several firms provide only limited results.

Only a few leading firms such as Toyota, General Electric, Dell and ABB have successfully harnessed intellectual capital to reap substantial benefits. For several others, it still remains a formidable challenge.

Leaders as Catalysts

Effective leadership continue to be the core of creation and implementation of winning strategies! Here, Prof. Peter Drucker's foremost dictum "to get ordinary people to do extraordinary things" assumes much importance. *Strategic learning* interlinked with *inspiring leadership* would help the global firm to perform the "alchemy of fostering regular innovation" – ensuring thereby a steady stream of innovative new products.

In-depth studies have revealed that leading firms such as: Siemens, General Electric, Sony, IBM, INTEL and 3-M have developed coherent processes to ensure that product innovation regularly occurs.

To be highly successful the global firm should formulate sensible, ambitious and systematic R and D efforts, which would involve the creation of an environment most conducive to perform creative work by talented engineers, designers, scientists and professionals within the global firm.

The CEO should make every possible endeavour to provide a pleasant framework which nurtures creativity and strategic thinking at all levels. To push the company to the forefront, the entire framework should be "infused with the spirit of discovery and innovation!"

Strategy for Excellence in a Borderless World

"As in the past the 'Impetus for Exploration' has come from 'business'. Once again, people with 'courage and curiosity' have discovered new ways of life, which are irrevocably changing the way of life on the old continents they left behind; the only difference is that the 'New Continent has no land'. It exists only in our "Collective Minds".

—Dr. Kenichi Ohmae
Director, McKinsey and Co.

I. THE GLOBALISATION PROCESS

The term Globalisation has been encapsulated in the phrase: 'think global and act local' and this has become the central theme of several major international organisations all over the world. As such 'Globalisation' has accentuated the trend towards an integrated worldwide economy. By and large, global forces have started influencing local lives in several countries/areas providing greater opportunities for enhanced cultural interaction.

Globalisation has made certain major changes as enumerated below:

   (i) Firms have started adopting long-term strategies which would help to establish their worldwide presence with standardised marketing programmes and practices.

(ii) They also resort to operational activities, which provide short term revenue generation to meet the local requirements.
(iii) The global firms have to cater for the needs of consumers in emerging new markets, with growing middle classes who are keen on buying more of global brands.
(iv) The following concepts, though may tend to overlap, have to be segregated and dealt with.
  (a) The term 'international' implies that a business is operating in more than one country.
  (b) The term 'multinational' implies a business, which operates in several countries.
  (c) The term 'global' indicates the worldwide scope of the organisation's business operations/capability to compete on a global scale.
(v) In a fast changing world the transport and communications network has brought about greater rapport among people and strategic alliances in the business communities.

Dramatic increases in the flow of goods and services, capital and information across borders have helped the rapid growth of technological infrastructure in different parts of the globe.
(vi) There is worldwide awareness to treat knowledge as an asset. Firms are being converted into 'learning organisations' which could suitably tailor their operations to meet the global market requirements.
(vii) Rapid technology improvements have brought about major changes by
  (a) accentuating innovative product developments
  (b) changes in production processes (eg: CAD/CAM) which are less labour intensive.

(viii) Several constraints imported by time and space have been dissolved on account of spectacular developments such as Internet facilities and satellite television.

In view of the above, the global firm has to provide greater focus on the customers, who have developed tastes for 'tailored products' in consonance with the latest trends in a highly competitive environment. Such a firm should be fast and flexible in approach and should be in a position to deliver customer value. Only organisations, which are able to offer 'scarce and distinctly excellent products', are able to survive amidst the growing competition.

On account of the 'Internet boom' after 1990's, an organisation which could act very swiftly with great flexibility alone could reach the forefront, and effectively tackle competition. Since geographical constraints have been removed by the Internet, organisations which are "instantaneously responsive" to the needs of the sophisticated customers are able to thrive. As a result, even large organisations segregate themselves into small groups to remain more responsive and flexible in their operations and their main focus is on 'fast changing customer requirements'.

On account of globalisation, several countries have become economically interdependent.

The onslaught of western (and especially American) culture exercises an all pervasive impact on all nations, and as a result they strive to preserve their 'distinctiveness' and 'core identity'.

Since traditional nation-states are being transformed into business networks, several countries such as France, Russia, Indonesia etc., make every possible attempt to project their 'distinct characteristics' to the maximum possible extent.

Use of Technology

In the context of rapid globalisation, only the organisation which perceives technology as a strategic vision would be able to gain considerable and critical competitive advantages.

Nowadays technological innovation is much related to 'information technology'. Recent trends indicate that management of information systems have become now far more 'available, affordable and sophisticated' than in previous decades. It is much acknowledged that by using information wisely, the modern organisation could reap substantial benefits, and thereby could maintain the 'lead' among several competing organisations.

The radical changes in technology could be attributed to micro chips, which are quite powerful and these are classified under 'artificial intelligence'.

Since knowledge is pouring in from various directions, the real challenge lies in how to identify and segregate the same and also put it to productive use.

A sustainable 'knowledge base' could be created only by:

(i) The correctness of information gathered

(ii) Time of availability of the required information

Highly successful organisations depend upon computers and information systems which convert knowledge into knowledge-base.

The modern CEO/manager could adopt multiple approaches to gain a clear picture of the function of the organisational set-up by using the information system on the following lines:

(i) Problems being confronted and the information required to solve these problems.

(ii) Appropriate decisions to be taken, and the information required for the same.

(iii) Identifying factors which are critical to achieve 'success' and ascertaining the type of information required.

(iv) Resources necessary to achieve certain 'end outputs' from the firm's activities.

Management Wizard-Michael Porter's Approach

The strategic planning techniques adopted by a global firm should take into account the 'threats' being posed by productive entrants.

Strategy for Excellence in a Borderless World 175

The following figure 8.1 provides a picture of the probable developments which a modern firm has to sensibly tackle.

Figure 8.1: Michael Porter's Competitive Forces-Model

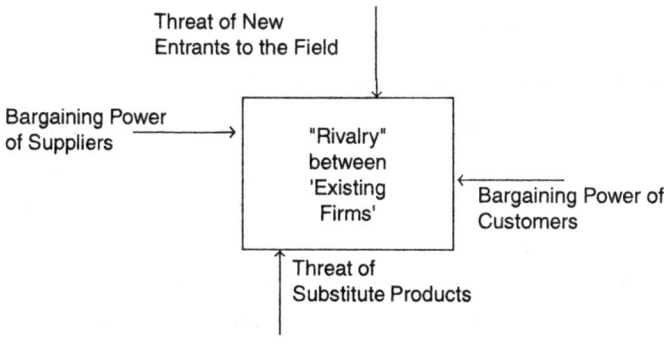

In this regard, Michael Porter has outlined certain strategies which could be advantageously adopted by the existing firm.

Strategies for Overcoming Possible Threats

(i) Overcoming 'bargaining power' of customers by installing terminals into customer's offices.
(ii) Reducing bargaining power of supplies by outlining/having alternative sources of supplies.
(iii) By setting up flexible manufacturing facilities, threats emanating from substitute products could be reduced.

Other Important Strategies (Generic)

Strategy (A): Performance of Value Activities at Lower Cost:
(i) Automating a manual process to reduce costs
(ii) Reducing considerably 'inventory-carrying costs'

Strategy (B): Differentiate own Product by Value Activities:
(i) For locking the customers by installing terminals in customer's offices.
(ii) Designating each methodology of approach.

Strategy (C): Endeavours to fill Niche Markets by Value Activities
  (i) Luxury passenger car buyers could be provided special considerations/plans.
  (ii) Home PC sales (for additional market) to network existing customers.

Inducting Other Improvements

'Survival' – in the highly competitive 21st century business environment is greatly dependent on the following factors:
  (i) Achieving technological superiority over rival firms/countries.
  (ii) Improving organisational capabilities.
  (iii) Concentrating on "high value-added operations".
  (iv) Need for speedy conversion of knowledge and technological competence into 'innovative products' to suit changing consumer preferences ensuring substantial profit margins.

In this regard, CEOs/managers must consult IT specialists to ascertain precisely which IT-investments would be found appropriate in facilitating 'value addition', besides improving performance of the firm as a whole.

Significant Changes Taking Place
  (i) More women would be inducted at different levels in the modern firms and may compete with men for high status jobs.
  (ii) Certain industries and markets are likely to expand rapidly, while a few others may contract.
  (iii) 'Technology' would be the foremost factor driving change. 'Interconnectedness' would be augmented, on account of improvements in transport links and communication facilities.

On account of revolutionary technologies, multifarious developments are taking place in respect of health care delivery.

Health care institutions and hospitals are undergoing radical modifications to facilitate better diagnosis and treatments.

Besides, in several advanced countries technological (biotechnology and science) advancements have helped to increase longevity and in providing specialised care to the disabled.

In this regard, it becomes highly essential to formulate a suitable mechanism that would ensure more meaningful work life for every employee in the modern organisation.

Operating Nationally and Internationally

In the global business environment, the pace and unpredictability of change is constantly increasing. In this regard it is needless to say that the CEO/global manager besides learning to build new knowledge should induct flexibility and adaptability in his organisational set-up in order to have competitive advantage. He should possess the 'global mindset' which is fully equipped with the potentially global scope of the business operations of his firm, with due regard for the flow of capital, technology, products and talent across borders.

He should understand that mere global presence alone would not prove helpful for gaining the required competitive advantage.

Besides, there is the imperative need to mobilise highly proficient people from different parts of the globe so as to establish a 'highly talented human resources base'. The organisation should also effectively handle 'production and logistics management' (How to augment main value adding activities connected with manufacturing'). There is also the need for organisational restructuring combined with planning-aimed at achieving a highly integrated global marketing strategy-covering product planning, design, development and so on.

Accordingly, these measures would involve formulating suitable financial strategies aimed at 'global competitiveness'.

Role Of Global Executives (Teams)

The 'global executives' should invariably be ready to tackle the following challenges:
- (i) Effectively managing cultural diversity and preventing conflicts.
- (ii) Access to sufficient communication covering different parts of the globe.
- (iii) Consensus in team work.
- (iv) Co-ordination/knowledge in handling:
    - (a) Teleconferencing
    - (b) Telephone
    - (c) E-mail
    - (d) Internet
    - (e) Desk top video conferencing (where there is no need to travel or gather at specific spots)
    - (f) Shared database indicating excellent record of team activities.

Requirements of a Global Player

It is an acknowledged fact that any enterprise keenly interested in becoming a global player should possess the following characteristics:
- (i) Ability to produce the right product.
- (ii) Products must be made available at the right time. Greater accent is laid on speed and accuracy.
- (iii) Quality of the products offered should ensure customer satisfaction. Several firms are keen on offering 'customer delight', with products exceeding 'customer expectations'.
- (iv) The enterprise should make every possible endeavour to be the lowest cost producer.

    All the plants attached to a global enterprise should be operating at 'zero defect'. 'Six sigma problem-solving methodology' is being implemented by several leading enterprises such as: Motorola, General Electric Co., Texas Instruments, ABB, Allied Signal, Ingersoll Rand Tool and so on.

The global firm should endeavour to be a 'customer driven company' by inculcating the ideals and ideologies enumerated in figure 8.2.

Figure 8.2: Ideals and Ideologies Necessary for a Global Firm (A customer driven company)

Total Productive Maintenance which would mean- Total Profit Management through... Total Perfect Manufacture through.... Total Productive Maintenance through..... Total People Management	
(Source)	Article of Manufacturing Excellence
	Suresh Krishna C and MD, Sundaram Fasteners

(v) During the 20th century, mass production techniques were highly popular, and firms mostly functioned by assuming away the actual customer requirements.

However, the 21st century milieu presents altogether a different scenario. Currently, customers are better educated and better informed than their parents or grand parents on account of revolutions in Information Technology. Their aspirations, tastes, fashions and fads are far beyond the modest expectations of earlier generations.

Since there is a quantum leap in the mindsets of customers as a whole, the modern firm could hope to gain competitive advantage only if it caters for the new attitudes/precise requirements of the new consumers! This involves a thorough grasp of the 'latest customer profiles' (categories) wherein the inmates' desires for achieving self fulfilment dominate. Therefore, only a firm which understands thoroughly customer preferences/ needs and provides a 'consistent level of quality service' would be able to gain distinct competitive advantage over rival firms.

The new strategy lays emphasis on 'mass customisation' (exceeding consumer expectations) and not on mass production techniques earlier in vogue.

II. DEVELOPING GLOBAL MINDSET AND SKILLS

To improve strategies in a fast changing environment, the CEO/managers of a global firm should necessarily possess a 'global mindset', which involves necessary cultural sensitivity and skills to enable the organisation to adapt suitably to foreign countries' overall environment. Besides eliminating old mindsets, the new multi-cultural CEO/managers should re-program their mental maps and constructs. They should endeavour to acquire 'multicultural competencies/skills' inclusive of foreign languages. They should also be able to 'envision trans-national opportunities' with great optimism.

In order to create a highly stimulating work climate, the entire organisational framework must be energised by the CEO's inspiring leadership.

The following figure 8.3 provides an useful sketch of the contrast between the traditional management mindset and the global mindset as presented by Dr. Stephen H. Rhinesmith in his *Managers Guide to Globalisation*.

Figure 8.3: Comparison of Traditional and Global Mindsets

Management	Traditional Mindset	Global Mindset
Strategy/Structure	Specialise Prioritise	'Drive' for Broader Picture Balance Contradictions
Corporate Culture	Manage Job Control Results	Engage Process Flow with Change
People	Manage Self Learn Domestically	Value-Diversity Learn-Globally

The CEO/manager having the global mindset essentially involves distinctly six definite attributes/personal characteristics. These are required to encounter and successfully tackle challenges in 'cross cultural interactions' in the global arena.

Important personal characteristics connected with the 'global mindset" and the implications are outlined in figure 8.4.

Figure 8.4: Personal Characteristics/Implications for Global Mindset

	Personal Characteristics	Implications
1	Highly Knowledgeable	Ability to manage competitive processes-both domestic and foreign
2	Highly Developed Analytical Ability	(a) Managing Global Matrix Organisations in 75 or 80 countries
		(b) Having 'Systems View' and understanding different levels of Vision/Mission/Strategy
3	New Strategic Vision	Honing strategic skills to adapt for rapidly changing Global Corporate Culture 'adding value to processes'
4	Flexibility to Adapt	Necessary technical expertise to respond to complexities quickly; Broad experience/ competency to take decisions based on experience
5	Sensitivity to Cultural Diversity	Fairly well developed self concept to tackle cross cultural challenges-involving broad range of people/cultures
6	Seeking Continuous Improvement Openly	To constantly review performance and induct improvements in the organisation/people

Source: Dr. Stephen Rhinesmith-Manager's Guide to Globalisation-1996

In regard to the most essential characteristics of a global mindset, Professor Jean Pierre Jeannet has aptly observed that the *global mindset* is a state of mind able to understand a business, an industry sector, or a particular market on a global basis.

The executive with a global mindset has the ability to see across multiple territories and focuses on commonalities across many markets rather than emphasising the differences among countries.

New strategies resulting from responses to new market opportunities are another part of this tool kit.

In today's globalising economy, the *'detection and identification'* of 'lead markets' is an important exercise. Soon after the conclusion of World War-II, the USA served as a lead market for many entrepreneurs in different parts of the globe. They established large firms simply by emulating US examples, without actively undertaking 'lead market analysis'.

Globalisation: Pathways and Leap Frogging

Once a firm gets the 'wake up' call, it may adopt different steps for the purpose of achieving global status after initially pursuing internationalisation process. The word 'global' represents not merely speed, but the 're-engineering of the appropriate framework' that causes several multi-domestic firms to struggle.

Some highly competent firms move directly to the most appropriate global level by 'leap-frogging' several stages.

'Leap Frogging Globalisation Pathways' has been highlighted in figure 8.5 given below:

Figure 8.5: Leap Frogging Globalisation Pathways

	Domestic	International	Multi-domestic	Global
Traditional Approach	□ →	□ →	□ →	□
Single Leap Frogging	□ →	□	→	□
Dual Leap Frogging	□	→		□
Skipping			□ →	□

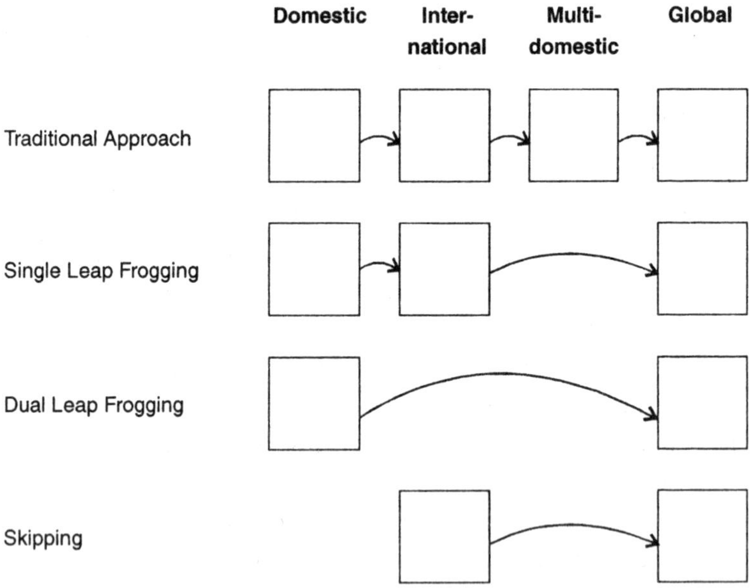

Source: Jean Pierre Jeannet: *Managing with a Global Mind-set*-(2000)

An Important Challenge to be Tackled

Some of the greatest challenges in the field of human resource development to be tackled by a modern organisation taking up 'Globalisation' pertains to the shift involved in converting country management from line to staff responsibilities.

In this regard, Dr. Moren and Dr. Riesenberger have provided details of global competencies, covering different mindsets and skills required by:

(a) A corporate CEO
(b) Subordinate general manager and staffers
(c) Other employees

Different categories of the staffers require a group of competencies selected from the twelve global competencies enumerated below:

1. Possessing a *global mindset*
2. Works on equal basis with persons from different backgrounds.
3. Having long-term orientation
4. Effectively facilitates organisational change
5. Creating learning systems
6. Motivating employees to reach high standards of excellence.
7. Negotiating and tackling conflicts in a 'collaborative mode'.
8. Managing skilfully the foreign deployment of expatriates.
9. Leading through effective participation in multi-cultural teams.
10. Thoroughly understanding one's own culture as well as national culture of others.
11. Making 'accurate profiles' of the organisational, national and different other cultures.
12. Avoiding cultural mistakes, behaviours in a manner that demonstrates knowledge and respect.

Identifying Competency: Competency Mapping

The process of identification involving enquiring each individual employee, who is currently performing a role to list the new tasks to be performed by him one by one, and identify the 'attitudes, knowledge and skills'-required to perform each of these. Later, a consolidated list of preferences/capabilities could be presented to a group of specialists, separately constituted for the purpose. The data could be edited and finalised so as to take suitable action.

Simulation programmes, role playing, interviewing and questionnaire methods could be employed during the process of identifying competencies.

The global firm, keen on making rapid strides of development, should certainly undertake a 'competency mapping exercise'. The global firm could be smartly managed by inducting the following approach:

1. The firm should have a well-formulated organisational structure.
2. The roles should be well-defined and the tasks and activities associated with each to be clearly spelt out.
3. The firm should have 'mapped' the 'competencies' most important for each role.
4. The consolidated data on competencies should be properly utilised for recruitment, promotion decision, performance improvement, placement and training needs of employees.

Such an approach would certainly facilitate the modern global organisation to identify suitable areas of improvement, and also help the staffers to 'enrich their specific competencies."

Besides, it would be found highly useful in making precise assessments of the 'intellectual, managerial, emotional and social competencies' of individuals and teams within the organisation.

III. MANAGING COMPLEXITY AND COMPETITIVENESS

Managing Uncertainty

Globalisation continues to be the economic buzzword of the times and uncertainty continues to play a crucial role. While firms and employees are lost in a haze, the CEOs/managers of leading firms focus greater attention on 'beating their competitors'. There is a growing tendency among managers to pay considerably less attention to technological and organisational strengths as well as their customers.

Since firms all over the world try to respond to a fast changing market place, firms are driven by the necessity to shuffle and reallocate the employees to business which would offer the highest returns.

'Competitiveness' is a blessing, on the one hand, with its concomitant uncertainty.

> (i) It is a blessing since people can enjoy a 'better quality of life' with access to the wonderfully fine products produced across the national boundaries.
>
> (ii) On the other hand, career management issue poses several complexities. The greatest problem is to make sure that the right jobs are allocated to the right and most suitable personnel besides allocation of labour to the correct place within a large firm in an unstable, fast-changing competitive environment.

Leading firms invariably kick off a series of initiatives such as: downsizing, de-layering, changing acceleration, six sigma and so on giving top priority to 'high value-added work'.

The new culture being created by the modern firm, therefore, should facilitate the requisite change only with due regard for ethical parameters. Only the organisational culture of the high performing companies focuses attention on each and every individual working in the firm on identifying and adapting to the latest trends in the market place.

In order to execute a successful strategy, the modern firm should present a suitable 'cultural vision' that would excite the imagination of, and provide the necessary psychological impetus to, the employees. The new culture should be formulated with meticulous care so as to make it responsive to the emerging competitive pressures.

The employees should be in touch with the market place/ trends and should thoroughly grasp the external challenges in order to ensure personal success, in addition to making substantial contributions for the healthy growth of the organisation. Again this depends on leadership qualities especially factors connected with promoting a 'fine cultural infrastructure', that would bring results in terms of profitability, greater market share, high adaptability to changing customer needs, and so on.

Multiple Objectives

Several challenges are presented to a global organisation and it should be capable of handling three important and conflicting objectives simultaneously:

(i) The absolute need to enhance efficiency in order to lower costs.

(ii) Adequate responsiveness to tackle a variety of 'customer requirements'

(iii) Perfect coordination sharing the 'best practices' on a global basis.

Geographic Scope of Operations

In the modern organisational set up, 'speed' and 'responsiveness' should be the foremost concerns in order to tackle effectively the new consumer demands and beat off competition.

To make meaningful contributions, the global managers would function in a framework where 'decentralised decision making' has been inducted.

The following figure 8.6 provides a snap-shot summary of the differences mostly characteristic of (i) a multinational company and (ii) a global firm.

Figure 8.6: Characteristics of a Multinational Company and a Global Firm

	Multinational Company	Global Firm
1	It establishes 'Mini Replicas' of its domestic business in many different countries/markets	It shares Resources on a global basis and has access to the best market with highest quality product and lower cost.
2	Local employees look after the Foreign Operations	It sheds its National Identity-Extremely sensitive to Global Changes/Trends; Constantly Scans Environment and Reorganises Resources and Capabilities. Regional and National boundaries are not potential to products, technologies and markets. (E.g. Coca-Cola, Gillette, Ford, Sony, Unilever etc.)
3	It resembles a Multi-domestic Organisation eg: 'Nestle'	

A global firm/corporation makes every possible endeavour to deliver the 'highest quality product' to the most profitable markets at the 'lowest possible cost'. It staves off bureaucratic controls and ensures 'flexibility' and 'rapid response to consumer needs'.

An useful diagrammatic comparison between a 'multinational company' and a 'global firm' as provided by management expert Stephen Rhinesmith is presented in figure 8.7.

In a global organisational framework invariably when a major re-organisation of decision making processes takes place, this would require changes in: corporate culture, systems, mindsets and skills of all managers in the entire organisation.

Figure 8.7: An Useful Comparison of Multinational and Global Organisations

Multinational Company	Global Firm
(A) One Centre	(A) Many Centres
(B) Hierarchical	(B) Network
(C) Rigid	(C) Organic
(D) Structure	(D) Process
(E) Boss/Subordinate Relation	(E) Interactive
(F) Chain Of Command	(F) Many Channels
(G) Information = 'Power'	(G) Information = "Resource"

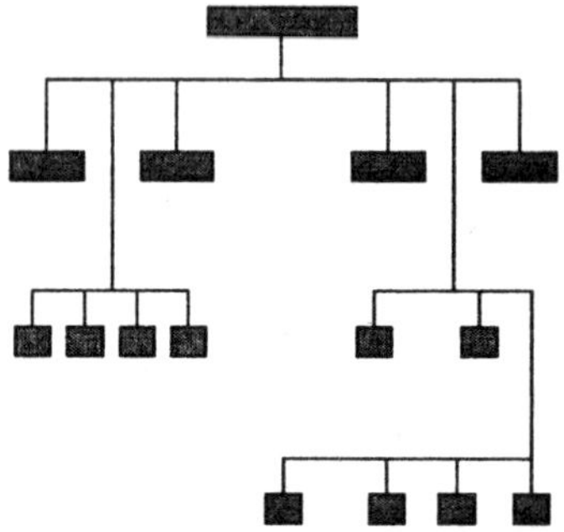

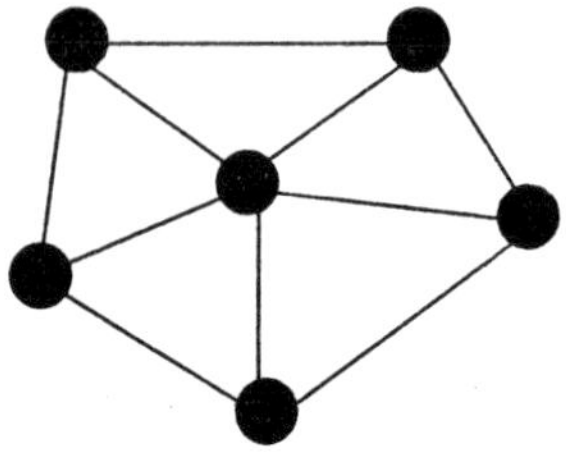

Source: Mr. Stephen Rhinesmith, 'Managers Guide To Globalisation'

Such an exercise involves:
 (i) The type of decisions that would be decentralised;
 (ii) The functional areas for decentralised approach;
 (iii) The kind of business structure most appropriate to function optimally.

Here, Dr. Stephen Rhinesmith has highlighted 8 different levels of 'functional integration' as enumerated below:

1. Globally across regions and product lines
2. Globally across regions within product lines
3. Internationally (international division only) across regions and product lines
4. Internationally within product lines
5. Regionally across product lines
6. Regionally within product lines

7. Locally across product lines
8. Locally within product lines

As for example, in level 8 – the Coca Cola company in order to suit the local preferences for sweetness/flavour etc., of its syrup, supplies slightly different flavours for:

  (*i*) Sweeter soft drinks for Asian countries

  (*ii*) Very mildly sweet-drinks for European countries

Coca Cola has also established R and D laboratories for each country to ascertain exactly the tastes of local consumers, and suitably arrange to supply the required taste, flavour as highly preferred.

'Global Norms' to be Followed

  (*i*) It has to be ensured (by managers and executives) that without over-riding individual cultural values in different countries, certain consensus has to be evolved by participation in specific forums through discussions/debates.

  (*ii*) Differences should be managed by 'synergy' (combined approach exceeding the sum of individual contributions) rather than focusing attention on one 'national perspective'.

  (*iii*) There is an absolute need for global managers to master and develop 'multi-cultural skills' in order to promote the development of a 'multi-cultural work force' within the organisation.

Naturally these measures strengthen vital organisational norms of behaviour, when encountered with complexities and contradictions in the global arena.

Flexibility in Approach

The global policy pursued by the global enterprise should have adequate flexibility in application by resolving certain inherent contradictions as given below:

(i) The firm has to be global and local at the same time.
(ii) The operations should cater for big and small needs, depending on demand conditions.
(iii) It should adopt a decentralised approach.

With a consolidated centralised 'reporting system', the global firm should be capable of delicately balancing and responding to the local business environment as well as the complexities to be tackled/economies and competencies required from the point of view of global economic scenario.

Decision-making

A competent global manager must be able to prevent obstacles by identifying the 'over-arching-problems' and 'executing neatly the global strategy'.

Fair decision-making involves the following conditions:
(a) The overseas head office should be familiar with branches/subsidiaries at local situations.
(b) Global strategy would be highly effective only when 'two way communication' (or feedback) exists.
(c) The decisions taken at the overseas head office should be fairly consistent in respect of branches/subsidiaries.
(d) The branches/subsidiaries should have the freedom to express their viewpoints and may suggest new options for betterment.

Besides, strategic decisions concerning international expansion should suitably take into account factors such as: market entry, new product development as well as production, sales and service, marketing and distribution channels.

Senior managers should be willing to apportion time for monitoring conflict management and empowering talented people to ensure that the best decisions are always made from the point of view of the healthy and rapid growth of the global organisation.

Suitable training programmes on subjects such as global trends, quality improvements, customer services etc., should be evolved to bring about rapid organisational changes.

While tackling problems on a global scale, constituting separate teams for different endeavours/projects would be highly instrumental in promoting 'creativity', 'organisational synergy' as well as 'innovations'.

The renowned General Electric Company provided wonderful insight into managing global change – "when a process nicely happens, the ability of managers to manage it is greatly enhanced".

The following 'analogy' illuminates the above idea:

"Every morning in Africa, when a gazelle wakes up, it knows that it must run 'faster' than the 'fastest lion', or it will be killed. Every morning, when a lion wakes up, it knows that it must run 'faster' than the 'slowest gazelle' or it will starve to death. So it does not matter whether you are a 'gazelle' or a 'lion', when the sun comes up you'd better be 'running'" (Hershock and Braun)

GALVANISING HUMAN RESOURCES

The term 'personnel manager', as traditionally used, has been enriched and modified to include practices such as employee empowerment and meticulous care in planning allocation of work, so as to designate the scientific name 'Human Resources Development' in recent years. It is needless to say that 'the human factor' is the most significant element which shapes the quality of output as well as the effectiveness of scientists in an industry. Therefore in 'industrial research and development' laboratories attached to a firm, the 'human resources development function' has been assigned a very crucial role.

A businessman keenly interested in having a competitive edge over his rival firms, and in earning substantial profits should judiciously select highly talented personnel, who would be able to respond and make use of the 'best practice bench marks' after undergoing necessary training.

'Human Resources Management' is greatly instrumental in projecting a better image of the firm, in addition to helping the organisation reach higher levels of performance standards reflected in terms of productivity, cost of production, accelerated growth and greater diversification of products.

When we analyse all the facets and ramifications of the concept 'Human Resources Management', we conclude that the individual employee associated with a modern firm should possess certain characteristics:

(i) He should be keenly interested/qualified to undergo requisite training meant for personnel development and in the interest of enhancing the long term benefits for the firm.

(ii) Increased knowledge of behavioural sciences and counselling in addition to having knowledge on various types of tests/measurements.

(iii) He should be responsive towards active collaboration in team work, and for higher level 'inter-personnel relationships'.

(iv) Besides having visible commitment for learning, he must have 'creativity' and 'innovative skills'.

The HRD department in a modern firm should design suitable training programmes for its employees which are highly appealing to them, keeping in view at the same time the 'mission and strategy' formulated by the firm to achieve spectacular results, and go to the forefront. It has to be ensured that the firm never drifts into a 'non-competitive position'. Besides, it should also periodically check up whether there is room for any 'technological obsolescence' within the firm. The most critical element for phenomenal success rests on 'strategic planning techniques'. The global enterprise should also inculcate the practice of adopting a long-term view for healthy growth, taking into account its present position, the stated objectives, the pace at which it should grow and the sensible steps to be taken to reach the goal.

For entering global markets at the outset, the firm should decide the potential effects of education/training needs-with due regard for the new language, new cultures, new business practices and so on. In this regard, the renowned 'Motorola Corporation' has achieved phenomenal success by establishing a separate university to impart training to its employees/staffers on a broad spectrum of topics at a cost of more than $ 170 million a year.

A fairly ideal model, which would facilitate effective HRD process in a modern firm has been presented in figure 8.8.

The following training methods could be advantageously employed:

(i) Demonstration
(ii) On-the-job training
(iii) Simulation models for training

Figure 8.8: An Ideal Model of HRD Process

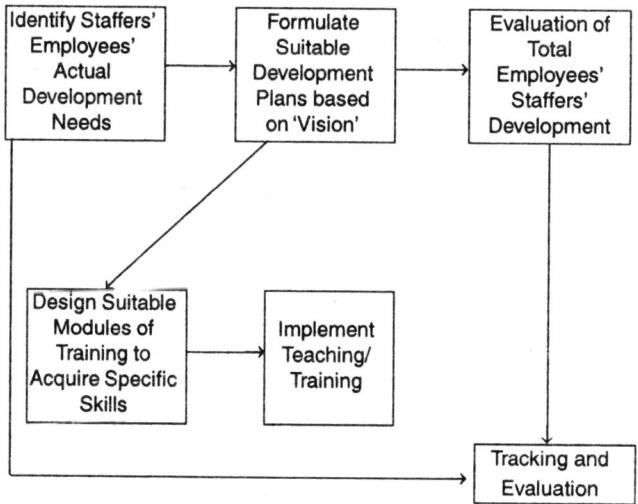

(iv) Class room methodology:
  (a) Lectures
  (b) Case studies
  (c) Meetings/Conferences/Brain-storming sessions
  (d) Role playing
  (e) Audio-visual programmes/Question-answer sessions
  (f) Providing suitable instructions.

John W Gardener has elegantly summed up the need for 'self development' in the following manner:

"If we are concerned with the shortage of 'talents' in our society, we must inevitably give attention to those who have never really explored "talents fully".

The Role of Human Motivation

The renowned Professor of Psychology, Abraham Maslow has provided an excellent picture through the 'hierarchy of needs' theory about the way in which people are motivated.

He has stated that "needs in lower categories" have to be satisfied, before "needs in the higher categories" could act as 'motivator'.

He postulated that human needs fall into five different categories-the lowest common to all living creatures as well, the highest needs unique to man!

It has been aptly stated "A violinist who is starving cannot be 'motivated' to play 'mozart', and a shop worker constantly working without a lunch-break is less productive in the afternoon than one with a lunch-break".

Prof. Maslow himself has narrated, "What conditions of work, what kinds of work, what kinds of management and what kinds of reward or pay will help human stature to grow healthy, to its fuller and fullest stature".

Further, he indicates "classical economic theory... provides an inadequate theory of 'human innovation'. It could be

Strategy for Excellence in a Borderless World 195

revolutionised by accepting the results of higher human needs including the 'impulse to self-actualisation' and the 'love for the highest values".

According to Prof. Maslow the five categories of needs form a 'hierarchy', and the individual position would be constantly shifting, when his or her need shifts:

(i) Physiological needs (at the base)
(ii) Safety needs
(iii) Social needs
(iv) Esteem needs
(v) Self-actualisation

These 'categories' are elegantly presented in figure 8.9.

Figure 8.9: Dr. Maslow's Hierarchy of Needs

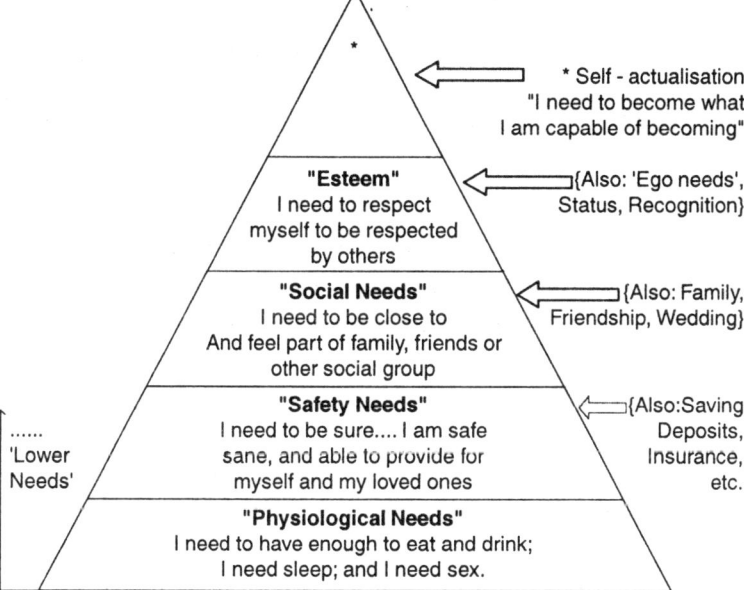

While elucidating self-actualisation, Prof. Maslow indicates: "A musician must make 'music', an artist must 'paint', a poet must 'write', if he is to be ultimately happy"... This is known as "self-actualisation".

When the participants "undertake a programme in an organisation", they see that 'change' or 'development' being offered by the 'programmes' really fit their "own primary purpose in life"... with their values, and the "realisation of their personal potential";... they will certainly appreciate such programmes, and show "intense commitment".

With democratic/religious slant, Prof. Maslow wrote: "One's only rival is one's own potentialities. One's only failure is failing to 'live up' to one's own possibilities. In this sense, every 'man' can be a 'king', and therefore be treated like a king."

ACHIEVING ORGANISATIONAL EXCELLENCE

Building New Capabilities – Global Approach

International growth presents a highly complex task. Factors such as local consumer preferences, currency fluctuation, government regulations applicable and cross-cultural management etc., should be taken into account before embarking upon building a 'strong global position'.

In a competitive global environment, it becomes absolutely essential at times to 'reconfigure' the entire business system with due regard to geographical/cultural factors. These are:

(i) Deciding the appropriate *location* for the manufacturing plant.
(ii) Deciding the *markets* in which the products are to be sold.
(iii) Planning to raise adequate *capital* along with sources.
(iv) How *inputs* could be arranged to flow into the plant from different parts of the globe.
(v) How to identify *talented managers* most suited for the business taken up.
(vi) Location of R and D laboratories and *product development centres*.

(vii) How to organise different locations for *excellent customer service.*

(viii) Deciding on *partnerships, acquisitions and alliances.*

Core Competencies

The concept of 'Core Competency' was first developed jointly by Dr. C K Prahalad and Dr. Gary Hamel, both renowned management gurus.

They have observed that 'Core Competencies' are:

(i) 'Collective learning' in the organisation, especially how to coordinate diverse production skills and integrate multiple streams of technologies.

(ii) Besides involvement and deep commitment and working across organisational boundaries, these competencies are always enhanced when shared.

(iii) A 'core competency' provides excellent access to a wide variety of markets.

(iv) A 'core competency' should make a significant contribution to the perceptible benefits sought by consumers.

(v) A 'core competency' cannot be easily imitated by the rival firms since it emanates from the 'harmonisation of highly complex technologies.'

These management gurus are of the opinion that if a modern firm could maintain world manufacturing dominance in core products, then it would have special ability to shape end-products.

The following examples provide ample testimony to the dominant position enjoyed by the following organisations:

Organisation	Special Expertise
1. Boeing	Aero Dynamics of Flight
2. Microsoft	Computer Science
3. Honda	Manufacturing Gasoline Engines
4. Sony	Turning out High Quality Miniaturised Products.

Besides, it is not enough for a particular firm to have simply one capability that is world class. In today's highly competitive business milieu, a firm should have several capabilities, classified as 'world class' and must always execute orders with meticulous care to reach greater heights.

Star Model on Core Competencies for Change in Performance

J R Galbraith, an eminent management thinker, has provided a 'star model' which would bring about significant changes in an "organisational performance" based on its core competencies.

The following five elements have to be examined and changed in order to be highly successful and to reach 'world class standards'. These are: *(a)* Strategy, *(b)* Structure, *(c)* Rewards, *(d)* Process, *(e)* People.

Figure 8.10: The Star Model: To Achieve Significant Change

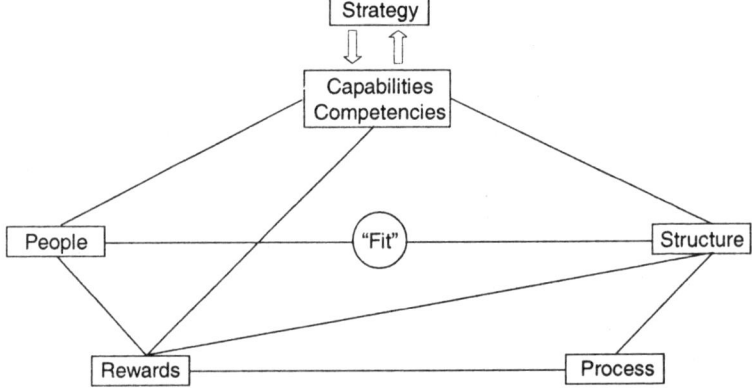

In The Star Model

(i) **Strategy:** Business strategy is the cornerstone design element. It outlines the organisational performance aimed at competencies and capabilities needed in the present business environment; besides how right products/services would be offered, potential customers identified, adequate financial resources earmarked and so on.

(ii) **Structure:** How teams are constituted, specific tasks are assigned and low 'critical decisions' made would be articulated.

(iii) **Rewards:** Rewards System would conform to the main strategy and correct behaviours would be rewarded. Attracting and retaining talented persons would receive importance.

(iv) **Processes:** The focus would be on 'correct strategic actions' and all the key processes would be well-motivated.

(v) **People:** Individuals possessing essential skills would be selected and recruited and 'the right people' would be imparted suitable training to achieve results.

Professor Michael Porter's 'Value Chain Approach' is a significant contribution on 'strategy formulation' and his Value Chain Model confers substantial advantages on the modern firm 'by creating value to the customers'.

Prof. Michael Porter's Value Chain Model has been presented in figure 8.11.

Figure 8.11: Prof. Michael Porter's "Value Chain Model"

Support Activities	Firm Infrastructure					'Margin'
	Human Resource Management					
	Technology Development					
	Procurement					
	Inbound Logistics	Operations	Outbound Logistics	Mktg. and Sales	Service	'Margin'

"Primary Activities"

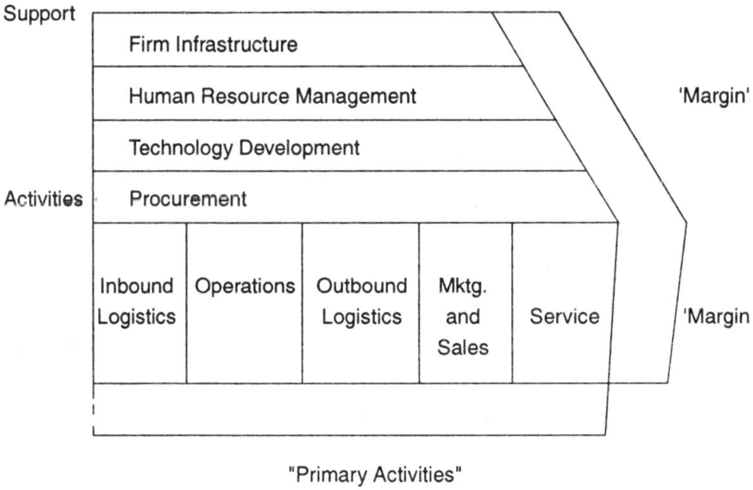

Prof. Michael Porter indicates that "a firm gains competitive advantage by performing these strategically important activities more cheaply or better than its competitors"

Primary Activities

Primary activities have been classified into five important categories:

(i) Inbound logistics:
They refer to activities associated with acquisition of materials used in the firm's product.

(ii) Operations:
They involve the transformation process of materials into finished goods

(iii) Outbound Logistics:
The storage and distribution of the finished goods to customers is included in this category.

(iv) Marketing and Sales:
They include the sale or inducement to buy the finished goods.

(v) Service:
The enhancement or maintenance of the value of finished goods, covering installations, repairs, etc., are included in service.

Support Activities

In regard to support activities, four generic categories are indicated:

(i) Firm Infrastructure:
Covers general management, finance/accounting.

(ii) Human Resource Management:
Involves activities connected with hiring, recruiting, training/development, etc.

(iii) Technology Development:
All activities connected with improving the quality of products/different production processes are covered.

(iv) Procurement:
  The function of procuring all inputs involved in the production process is included in this type of support activity. In the above 'model', margin would constitute the difference between the total value and the cost of performing the value activities involved!

  When a consumer is convinced of the superiority of the product, he would be willing to pay 'premium price'. However, the premium depends upon the consumer's attitude.

Outsourcing

'Outsourcing' takes place in a situation where a firm opts for purchase of goods and services from external sources/vendors, when these could be manufactured by the firm itself.

Firms prefer outsourcing in respect of activities such as:
(i) Information Technology
(ii) Human Resources (Talented Personnel)

Outsourcing has adequate flexibility since it could be arranged on a contract basis, and the outsourcing clients have the option to put an end to the contract at a suitable time.

Outsourcing is undertaken mainly with the following ends in view:
(i) To improve efficiency and effectiveness within the firm.
(ii) Effecting reduction in fixed costs
(iii) Pursuit of labour flexibility
(iv) For effecting improvements in products/services.
(v) Access to new skills/adaptation to technological changes.
(vi) To facilitate faster development of new products and services.

However, it should not endanger job security and opportunities for advancement within the organisation.

Customer Focus

Successful firms constantly develop new products on a regular basis. Since the demands of the market place constantly change, the vast majority of the products needs to be continuously improved and updated to enable the necessary volumes of sales to justify their profitable manufacture.

'Product development' decisions will have to take into account:

(i) Latest market trends
(ii) Competitive activities
(iii) Customer requirements
(iv) Size/priority etc.

In addition, there are certain factors internal to the organisation which exercise influence.

These are: strategic goals, long-term positioning, product portfolio financial resources available, etc.

In regard to new product developments, industrial manufacturers invariably concentrate on three key areas.

(i) Organisational structure/culture
(ii) Appropriate technology
(iii) Skills required with creativity

The manufacturer is keen on employing the most appropriate technology which would truly make the new product development: 'both better and faster', besides facilitating the reduction of costs in every respect. Only by adopting the enabling technology which would help to make the manufacturing company more competitive, the firm could hope to get substantial benefits/profits in both domestic and international markets. It is utmost essential for a global firm 'to reach very early to the market with a well-designed product', so that it could enhance its market share and reap great rewards.

Concurrent Engineering

Concurrent engineering methodology could be advantageously used since it involves the appropriate application of process improvement tools and techniques combined with those of modern engineering such as: Computer Aided Design (CAD) engineering and manufacturing.

Team work is the most essential ingredient in concurrent engineering-and several well-qualified teams participate to ensure that the products supplied by the firm provide the optimum satisfaction to the customers' needs and expectations.

Concurrent engineering techniques ensure the following characteristics:

(i) Marvellous designs with excellent performance standards
(ii) High reliability/maintainability
(iii) Lower costs with faster deliveries.
(iv) A rapid development production cycle: ensuring 'product leadership' with substantial profit margins.
(v) Computers help engineers/experts to see and test new concepts on the basis of Computer Aided Design data.
(vi) 'New product development' is not the sole domain of engineering department. It becomes a joint venture, where all capable people participate to turn out the 'best total quality products'.

Re-Engineering

No doubt the 'technology revolution' has thrown open several options before manufacturers to select suitable technologies. However, new technological solutions to improve productivity and quality would be found really helpful only when such technological changes could be harmoniously integrated with a well-formulated new strategy evolved for the specific needs of the modern firm.

Prof. Michael Hammer and Prof. Champy – renowned experts in the field of 're-engineering' – have stated, "The fundamental rethinking and radical redesign of business processes achieve dramatic improvements in critical contemporary measures such as: *'cost, quality, service* and *speed.'*!

Business process re-engineering was implemented with considerable success by firms such as: Kodak, the Hall Mark Card Co., etc.

Re-engineering has been instrumental in 'streamlining the activities' of several organisations to suit the specific modern requirements.

Just in Time

It was introduced in 1970 by the Japanese. It provided a 'radically new approach to manufacturing process.' It was able to 'cut waste altogether' by supplying components/parts only as sand when the process required! Besides JIT eliminated holding buffer stocks and accumulated inventory.

The system allows a variety of models to be produced 'on the same assembly line simultaneously' instead of concentrating on one 'single model' at a time.

In JIT, Kanban (Cards) were essentially utilised and here a card would be sent to 're-order a standard quantity of parts and components' required during a manufacturing process.

Prof. Russel and Prof. Taylor, have aptly described JIT as "If you produce only when you need it, then there is no room for error. For JIT to work, many fundamental elements must be in place – steady production, flexible resources, extremely high quality, no machine break-downs, reliable suppliers, quick machine setups and lot of discipline to maintain the other elements."

JIT has occupied the centre of the 'total quality movement', and of the 'flexible manufacturing techniques' that are the essence of 'lean production.'

Several leading firms such as: Toyota, General Electric, Hewlett Packard, Texas-Instruments, Omark, Westing House etc., have successfully experimented with it.

Bench marking

'Benchmarking', the analytical tool meant for improving organisations' processes, products and services is utilised by several organisations on a global scale.

'Benchmarking' could be defined as:

"Making useful comparisons of products processes, methods, services with the 'best practices' found in other organisations, and sensibly adopting them as quality improvement projects by any aspiring firm."

Foremost objectives

Benchmarking has specific focus on certain factors:
- (a) Analysing 'operations' thoroughly to identify strengths/ weaknesses in the existing processes at present.
- (b) Identifying the leading firms' (in the product line) 'best practices' and suitably adopting them.
- (c) Making every possible endeavour to become 'the best' among the topmost (best) firms.

Several renowned companies had undertaken this exercise with phenomenal success. These firms are: 'Xerox' (which won 'Malcolm-Baldridge Quality Award') AT and T, Motorola, Texas Instruments, General Electric, Digital Equipment Corporation, etc.

Thus, benchmarking is a comparison process which continuously identifies 'the best business practice' anywhere in the world, and utilises these for the benefit of an aspiring organisation.

VI. EXCEPTIONALLY SMART LEADERSHIP

All business organisations are experiencing unprecedented changes in the 21st century milieu, and the future still appears uncertain with moments of great challenges.

Several leaders are overwhelmed by complex responsibilities and overloaded by voluminous information which require constant review and segregation.

Dr. Warren Bennis has aptly observed, "There is general agreement about the 'qualities of leadership', but there are no 'specific formulas' for growing leaders".

He has further stated that derailment at top leadership levels take place not on account of lack of business literacy or conceptual skills, but mainly because of 'lapses of judgement' and 'questions of character'.

Quite paradoxically 'good judgement' and 'character' combined with an 'integrated personality' are mostly ignored! CEOs/executives who facilitate learning should, therefore, serve as wonderfully fine 'role models', in order to have positive impact on subordinate staffers/employees.

The following characteristics are mostly found in great leaders.

(i) The leader or CEO should be capable of 'transforming the mind-set of staffers' for achieving outstanding performance.

(ii) He should possess 'healthy aspirations' and should be receptive to 'new ideas' and learn from experience.

(iii) He should be in a position to identify and unleash the creativity, know-how and proactivity found among his employees.

(iv) His foremost aim should be to achieve competitive advantage over rival firms, and in this regard he should be well-versed in the art to orchestrate multifarious talents and generate intellectual capital. He should be willing to announce adequate rewards for achieving excellence/outstanding performance among his employees as well as dedicated mentors and coaches.

(v) He should constantly get authoritative feedback information both (positive and negative) in order to make effective corrections.

As could be seen from figure 8.12, one factor that has a direct causal relationship with financial performance is the measure of quality and client relationships.

Figure 8.12: The Causal Model of David Maister

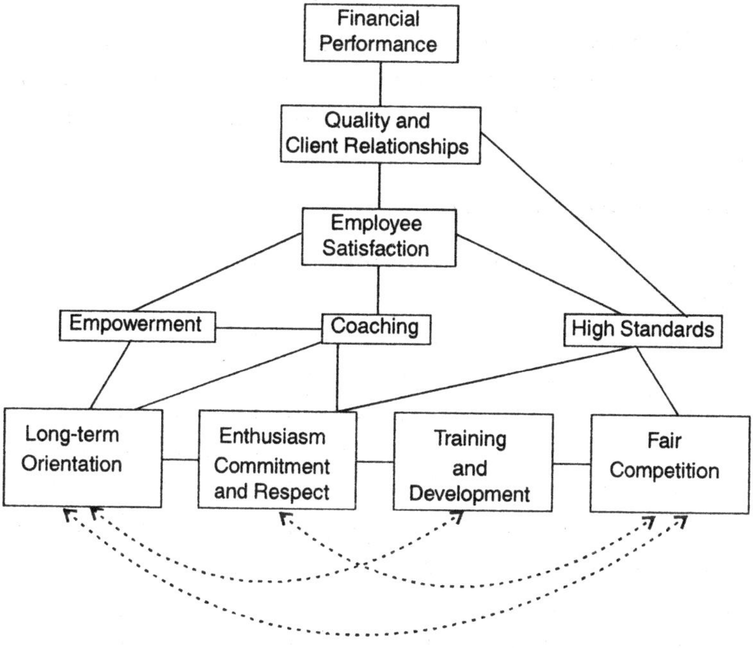

```
                    ┌──────────────┐
                    │  Financial   │
                    │ Performance  │
                    └──────────────┘
                           │
                ┌──────────────────────┐
                │     Quality and      │
                │ Client Relationships │
                └──────────────────────┘
                           │
                    ┌──────────────┐
                    │   Employee   │
                    │ Satisfaction │
                    └──────────────┘
     ┌─────────────┬─────────┴──────┬──────────────┐
┌───────────┐  ┌──────────┐              ┌───────────────┐
│Empowerment│  │ Coaching │              │High Standards │
└───────────┘  └──────────┘              └───────────────┘
┌──────────┐ ┌────────────┐ ┌──────────┐ ┌──────────┐
│Long-term │ │ Enthusiasm │ │ Training │ │   Fair   │
│Orientation│ │ Commitment │ │   and    │ │Competition│
│          │ │ and Respect│ │Development│ │          │
└──────────┘ └────────────┘ └──────────┘ └──────────┘
```

The above named factor is made out of certain characteristics.

(i) The firm should make its clients feel as if they are very important.

(ii) It should periodically keep its clients informed on issues affecting the business.

(iii) The firm should have real commitment to high levels of client service.

(iv) The quality of work performed for clients should be consistently high.

(v) The firm should endeavour to build long-term client relationships!

As such 'measure of quality and client relationships' would constitute 'characteristics' enumerated above.

In addition, from the point of view of satisfying clients, the firm should try to undertake whatever jobs are entrusted to it, and also endeavour to resolve client problems when they occur. Besides, the clients' interests should be given priority attention even overriding the interests of the staffers within the organisation.

Charismatic Leaders

Charismatic leaders are viewed as the 'magic elixir' to cure several organisational woes such as:

(i) Turning round ailing corporations

(ii) Revitalising old bureaucracies

(iii) Launching new enterprises with innovative ideas

'Charismatic leaders' are universally celebrated as heroes of management because they powerfully communicate a 'compelling 'vision' of the future, propound innumerable 'creative ideas', and promote their beliefs with 'boundless energy' and 'enthusiasm'. Charismatic leaders with ethical principles focus mostly on developing people with whom they interact to higher levels of ability, morality and motivation.

Leadership and Management

Prof. Kotter, while contrasting 'management' and 'leadership' indicates that as such an organisation will need both strong leadership and strong management.

The following figure 8.13 projects essential features of the two contrasting styles: 'management and leadership'.

Strategy for Excellence in a Borderless World

Figure 8.13: 'Management' Vs 'Leadership'

	Management	Leadership
1	**Planning** Setting Targets, Identifying steps to achieve Goals; Allocation of Resources	**Envisioning** Setting 'Direction' and Creating a 'Vision' for the Future along with 'Strategies'
2	**Organising** Creating Structure/Jobs Identification/Staffing Communicating Plan of action	**Aligning People** Communicating 'Vision' and Marshalling 'Support', Empowering them with Sense of 'Direction'
3	**Controlling/Problem Solving** Instals 'Control System' to detect Deviations from 'Plans'; Ensures Completing Routine Jobs Successfully.	**Motivating/Inspiring** Energising People with 'Greater Involvement' supporting 'Employees Efforts', Recognising and Rewarding their Successes, Strengthening Informal Relationships.

Decision-making

Operational business decisions could be successfully achieved by:
- *(i)* Effective management of knowledge/information base.
- *(ii)* Fostering creativity and innovation.
- *(iii)* Focus on continuous improvement.
- *(iv)* Fixing operational decisions suitably within the overall strategy.
- *(v)* Suitably attuning the corporate culture.
- *(vi)* Building a pleasant environment for effective interaction without any hostility.

Fostering Innovation and Creativity

Professor Kenichi Ohmae in his classic work: *The Mind of the Strategist* provides excellent insights into 'creativity'. He observes, "Most Important, I believe, we need to cultivate 'three interrelated conditions' (i) an initial charge (ii) directional antennae and (iii) a capacity to tolerate 'static'; call it what you will, vision, focus, inner-drive–the 'initial charge' must be there."

Besides, he states:

"If the initial charge provides the creative impetus, directional antennae are required to recognise phenomena... which, as the saying goes, are in the 'air'. These 'antennae' are the component in the creative process, that uncovers and selects among a 'welter of facts' and existing conditions, *'potentially profitable ideas'* that were always there, but were 'visible' only to eyes not blinded by habit.

"Creative concepts' often have a 'disruptive' as well as a 'constructive' aspect. 'Successful innovative strategists' should have the static tolerance component of creativity... which covers the will to cope with criticism, hostility and even derision."

In leading firms, inducting radical changes and re-engineering would facilitate and foster innovations. Every employee should be a learner and every occasion should be a learning opportunity. Besides, authoritative knowledge/data and experience of others should be appropriately utilised and distributed in order to shape 'a global learning organisation'. It should also be ensured that innovative projects are realistic and should be tested for practical implications (implementation).

On Generic Characteristics

Several management experts have expressed the view that 'leaders need to be born with a set of genetic characteristics' which provide the raw materials from which 'leadership' could be well-nurtured. Experienced management experts are able to segregate those who possess the 'gene of leadership'.

It is believed that 'leadership potential and style of behaviour' start at a very young age!

Certain characteristics such as: high intellectual ability, character, charisma and attractive personality are also indicative of 'leadership potentialities.'

Split Brain Theory and Effective Management

Dr. Roger Sperry, the Nobel Prize winner, developed the Split Brain Theory. According to him, the 'right' and 'left' hemispheres have different overlapping functions.

Scientists, especially neurologists have found out that the 'left hemisphere' controls movements of the body's right side, while the 'right hemisphere' controls movements of the left.

(a) Right Hemisphere

It is the source of imaginative artistic thinking.

Mostly artists, athletes, politicians may have better developed 'right hemispheric process'. In schools, right brain processes are given less importance.

Esoteric psychologies (Eastern) such as: zen, yoga and sufism have focused on 'right hemispheric consciousness'.

(b) Left Hemisphere

It is known for its logical analysis and analytical thinking. Mode of operations is linear and information is processed sequentially one bit after another; linear faculty is language.

Most lawyers, planners, accountants have developed more of 'left hemisphere' thinking processes.

Western psychology (in contrast to East) is concerned more with 'left hemispheric consciousness' and logical thought.

Figure 8.14: Split Brain Theory-Essential Features

Left Hemisphere	Right Hemisphere
1 Logic	1 Intuition
2 Sequence	2 Images
3 Verbal	3 Visual
4 Linear	4 Spatial
5 Analytical	5 Creative
6 Reasoning	6 Holistic
7 Explicit	7 Colours
8 Calculation	8 Emotions

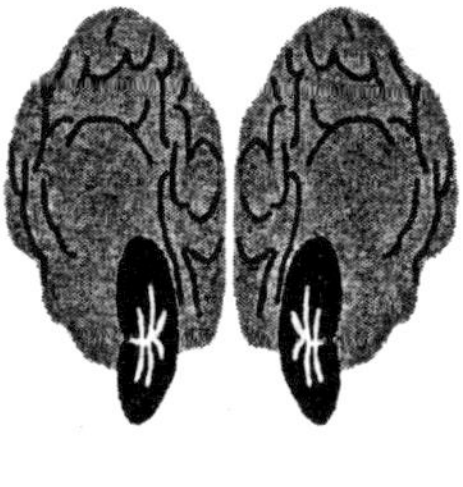

Source: Prof. John Adair: Effective Innovation

Prof. John Adair states that there are three meta functions stemming from the 'Evolutionary Theory of Human Mind', in response to its environment namely

(i) Analysing
(ii) Synthesising
(iii) Valuing

Research Findings on Management Processes

(i) Some people are capable of mastering certain mental activities, and they are incapable of mastering others.
(ii) Several creative thinkers find it difficult to comprehend a balance sheet, while accountants have no understanding on product designs.
(iii) Planners/ordinary management scientists revel in a systematic, well-ordered world (left hemisphere), and indeed they show little appreciation for more relational, holistic processes (right hemisphere).

Some important conclusions arrived at by research findings are enumerated below:

(i) The important processes of managing an organisation rely, to a considerable extent, on the faculties identified in the brain's right hemisphere.
(ii) Research findings indicate that senior management processes (involving 'feel' for the situation: not merely decision based on 'concrete data') belong to the 'right hemispheric thinking'. A person having good intuition means that he or she has good 'implicit models' in his or her 'right hemisphere'.
(iii) As regards strategy making in organisations, it will not turn out to be regular, continuous and systematic. 'Strategy' represents the mediating force between 'dynamic environment' and a 'stable operating system'. The burden to cope falls on the manager specifically on his or her mental processes – 'intuitional and

experimental' – that can deal with irregular inputs from the environment.

(iv) Research scientists mostly agree that the important 'policy level processes' required to manage an organisation rely, to a considerable extent, on the faculties identified with the brain's right hemisphere.

(v) However, the wonderful intuitive powers associated with the right hemisphere are obviously useless without the faculties of the left.

(vi) Several management schools with closed systems teaching theories of mathematics, economics and psychology with little exposure to the reality of organisational life, management ideas etc., cannot achieve much.

Therefore, there is need for a 'new balance' – the best human brains can achieve between 'analytic' and 'intuitive'.

(vii) Research scientists mostly agree that truly accomplished and outstanding managers are no doubt the ones who can couple effective processes of the 'right' (intuition, hunch, synthesis) with effective processes of the 'left' (articulateness, logic and analysis).

Need for Tackling Leadership Shortage

David Whitman, CEO of Whirlpool Corporation has aptly observed:

"The thing that wakes me up in the middle of the night is not what might happen to the economy or what our competitors might do next; It is worrying about whether we have the 'leadership capacity and talent' to implement the 'new and more complex' *global strategies*".

In the new millennium, every organisation faces the same daunting dilemma – 'the demand for leadership talent' far outstrips the 'actual supply'.

Need for Drastic Changes:
Accelerating Executive Talent

In an aspiring organisation, instead of hand-picking one or two-people for executive position (TOP), a group of 'high potential candidates' could be scientifically and rapidly groomed by constituting a 'separate pool' for executive jobs in general.

The 'pool members' should attend specific university programmes (suitably tailored) and 'in-house programmes' (involving action-learning exercises), which would help them to focus on skills and knowledge required for top positions and for effectively managing global problems. This would ensure grooming of highly capable 'tomorrow's leaders'.

VII. TOWARDS WORLD CLASS MANUFACTURING

The concept 'world class manufacturing' would mean: manufacturing and offering 'the best quality product' in the field which can successfully compete with the best in the world. Some products have a timeless appeal to a significant section of buyers.

Significant features of world class product would be:
 (i) Superior product design and performance standards
 (ii) High quality customised products
 (iii) Cost quality improvements through concurrent engineering.
 (iv) Capability to introduce innovative designs at a faster pace than competitors.
 (v) Shorter lead times and highly reliable delivery.
 (vi) Sophisticated products having an edge over others in foreign markets.

Prof. Maskell has stated that world class manufacturing as such broadly covers focus on quality, just in time (JIT) production techniques, work force management and flexibility in meeting customer requirements.

Since several developing countries are functioning in a highly turbulent, dynamic, and complex environment, they have taken up a process of 'restructuring their economies' to accentuate competition, and achieve 'suitable integration with global markets'.

To be globally competitive, several leading firms in developing countries are making ardent endeavours to achieve 'world class manufacture status'.

'Manufacturing excellence' ultimately depends upon 'manufacturing technology' which would constitute two important aspects:

(i) Technology of the product
(ii) Technology of the process.

Technology of the Product

To provide utmost satisfaction to consumers and enjoy sizeable market share, the product turned out should have 'high performance standards' with well-built features and sophistication. It should be 'user friendly' also. Rudolbh G Boznak states that 'stability' could be maintained when a manufactured product's physical definition consistently equals its documented definition throughout its 'life-cycle'.

Technology of the Process

The manufacturing framework should inculcate the cost reduction philosophy with 'elimination of wastage in toto".

The term 'agile manufacturing' should be thoroughly understood, and implemented.

The "Agile Manufacturing Model", pioneered and presented by Toyota Motor Company assures to supply 'a car designed as per customer's requirements' in 3 days' duration.

The basic idea in such a production system is 'to produce the kinds of units required at the time needed, in the quantities preferred'. Thus the unnecessary intermediate and finished product inventories are eliminated.

Total Quality Management (TQM) practised by a global firm should be essentially customer centric, and 'quality' is perceived as the responsibility of the entire organisation.

For 'world class manufacturing' to be really successful 'both product technology and process technology' should go hand in hand with judicious combinations.

Three kinds of 'flexibility' assume great significance in this context.

   (i) Product-mix flexibility
   (ii) Design flexibility
   (iii) Volume flexibility.

To tackle effectively the rapid innovations taking place, the organisations aspiring 'world class status' should 'decentralise decision making', broadly define jobs, develop a few procedures and subjectively 'evaluate performance'.

Six Sigma Production

Azim Premji of 'Wipro' has asserted that "quality like integrity is not negotiable".

'Six sigma' should constitute the 'way business is being managed' by a global firm.... in order to be highly successful.

Success of Six Sigma

   (i) Top management should constantly communicate its belief in Six Sigma and it should be 'centrally orchestrated'.
   (ii) Besides making substantial investments for the 'requisite training and tools' at least 15 to 20% of working time should be devoted for 'six sigma'.
   (iii) A sensible balance between financial benefits and customer satisfaction to be achieved.
   (iv) The most crucial element for the success of a 'six sigma' project is adequate investment in training resources.

'Defects' should be totally eliminated and the firm must ensure a 'top down approach' to 'six sigma approach' in order that 'initial resistance' by employees could be effectively tackled!

(v) There is need for fixing up a time frame of 6 to 8 months for small projects. To achieve significant results as a whole, a firm may require at least 2 to 3 years duration.

VIII. STRATEGY FOR A BORDERLESS WORLD

It is universally acknowledged by management experts that 'global competition' and 'global opportunities' on a vast scale have become all pervasive. Therefore, aspiring firms should think 'internationally' and formulate 'plans globally' in order to be really successful. The radius of competition for all firms has acquired a 'global dimension'.

An organisation's 'responsiveness' to changing market conditions (with smart management) has replaced the old myth 'bigness confers advantages'. Mere global presence alone will not ensure success for a global firm. The aspiring global firm, besides leveraging infrastructure, should be capable of mobilising an excellent 'global talent pool'.

Besides developing a 'global mindset', a firm which aspires for 'world class status' should make an in-depth analysis of the following factors:

(i) At the demand side, what most customers are likely to want in future and the factors, which would influence their purchase decisions should be thoroughly grasped.

(ii) At the supply side, how well the firm is able to tackle these wants, and how the rival firms are functioning are to be analysed.

The global firm, before formulating a 'world class action plan' meant for deciding the future development of the firm, should at the outset arrange to hold detailed discussions with the members of the top management team constituted for the purpose.

The members of the team should discuss multifarious facets in all ramification, and suitably formulate a 'consolidated strategy' without any internal contradiction of opinions within the firm.

Before developing the world class strategy, decisions pertaining to the following issues must be thoroughly worked out:

Where the Firm 'Stands' Now ?

   (i) What are different products and services being made available at present?

  (ii) Who are the firm's customers ?

 (iii) What about the firm's 'performance standards' compared to those of competitors?

 (iv) What are the unique strengths of the firm which would help to build a better future?

  (v) What are the weaknesses which have to be effectively handled?

 (vi) What is the current economic position in different parts of the globe in which the firm is interested, in effecting sales and the regulatory measures/political changes at home and abroad?

Such a review is called 'SWOT Analysis' which helps to understand the 'strengths, weaknesses, opportunities and threats' for the aspiring firm.

The Firm's Mission and Where it Wants to be ?

The firm should be able to clearly segregate the products/services, it could offer to the target markets in the short term (within 2 years) as well as those on a 'long term basis'.

It should be capable of adapting itself within the shortest span to an emerging situation 'when the competitive landscape changes' and should formulate an 'excellent blue print' for the firm's growth on healthy lines.

What Action is Required?

The firm should formulate dynamic strategic plans to tackle effectively the 'multi-dimensional phenomenon of globalisation', encompassing globalisation of supply chain, capital base, and that of mindsets.

Factors to be Considered When Competing in Global Markets

Product Differentiation

The global firm should create a favourable impression on the minds of foreign customers that its products/services are 'unique', having superior features than those of foreign competitors.

Market Dominance

The global firm should be in a position to offer a 'range of excellent quality products' at comparatively lower prices than those of the rival firms; otherwise the prices may be in line with those competing firms. It should endeavour to enjoy a sizable market share.

Factors which assume great importance in this context are:

(i) Good access to raw materials at competitive prices.
(ii) The designs formulated should be highly appealing
(iii) If quite a good number of 'products' could be made available on a global basis, the firm could enjoy advantages on account of a 'high degree of synergy'.

The global firm should also install 'state of the art', highly efficient machinery and equipment, in addition to employing 'highly trained/talented work force' which is greatly motivated by 'continuous improvements' principle.

Carving out a Niche Strategy

The customers should keenly opt for the firms' products on account of their sophisticated, well-designed and user-friendly features offered at reasonable prices.

A 'Niche Strategy' could be formulated on the basis of product differentiation or costs, and these products should be made available to consumers at the time stipulated.

Use of Benchmarking

Competitive benchmarking technique could be used as an important tool for facilitating resounding success.

Factors such as (i) reasonable price (ii) delivery in time (iii) reliability (iv) ease of reproduction (v) immediate availability of spare parts/components, would certainly reinforce the firm's position to offer an 'excellent range of world class products'.

Constant Review of Mission and Strategy

Besides constantly reviewing the 'short-term' and 'long-term' objectives of the firm, and reformulating them as per the current business scenario, the firm should endeavour to:

  (i) Expand into different geographical areas instead of depending on a few.
  (ii) Planning greater flexibility into the total capacity.
  (iii) Making endeavours to stimulate 'greater demand' for the firm's products.

Constituting a Think tank

Since 'decision-making' is a highly complex task in the modern dynamic business milieu, the CEO/professional manager should listen to the viewpoints of 'other talented staffers' before decision-making on important issues. There should be a 'think tank' which would act as a 'problem solving mechanism'.

Such an interaction would expand the intellectual horizons of every one in the firm covering: world economics, international law, techniques of business management, market place needs and so on.

Transformational Changes – Becoming Global

Managers of leading firms all over the world endeavour to implement 'radical and transformational changes' in firms, by throwing off their old management models in order to cater for global needs.

One foremost truth is perceptible: "No Change will occur until people change".

The actual voyage from a structural hierarchy to a 'self-renewing individualised (global) corporation', though found exciting is likely to be long and painful.

Here it is analogous to a caterpillar transforming itself into a beautiful butterfly. In the actual process, the caterpillar may find it intensely unpleasant by becoming blind, with legs falling off, and the body may experience deep cuts, when beautiful wings are emerging.

Management experts agree that the transformation into a global organisation is quite a painful process as per the above example; but the firms that succeed will certainly develop new behaviours most essential to take flight as 'successful global corporations'.

Functioning in a Global Multi-Cultural Environment

CEO's/executives would face several challenges inclusive of better ways of managing 'resources – human, financial, physical and informational.' To stay ahead, effective changes are required in terms of product development and induction of technologies.

Same firms may opt for making people work smarter (by training) so as to meet customer requirements in a better manner.

To achieve 'global competitiveness' and resounding success, the modern firm essentially requires 'multi-cultural managers' (inclusive of CEO/executives) who possess the innate skills and cultural sensitivity to push forward the organisation in a foreign country's overall business environment.

These 'multi-cultural managers' should possess certain essential characteristics:

(i) They should have new mindshifts with ability to think beyond local perceptions in a positive manner.

(ii) Besides 're-programming their mental maps', they should readily adapt to 'new/unusual lifestyles' and circumstances.

(iii) Over and above being well-versed in foreign languages, they should also acquire 'multi-cultural competencies'.

(iv) They should tackle comfortably without any inhibition clients/people from different disciplines, backgrounds and genders.

(v) They should be able to render really worthwhile assistance to the organisation operating in a different country. They should be able to implement the strategies formulated by the global firm to which they belong.

The following figure 8.15 provides essential details of cultural values, mostly appealing to the U.S., Japanese and Arabic countries.

Figure 8.15: Cultural Contrasts/Values Appealing

	'United States'	'Japanese'	'Arabic'
1	Freedom	Belonging	Family Security
2	Independence	Group Harmony	Family Harmony
3	Self-Reliance	Collectiveness	Parental Guidance
4	Individualism	Age/Seniority	Age
5	Efficiency	Group Consensus	Devotion
6	Directness	Quality	Hospitality
7	Aggressiveness	Interpersonal	Formal Admiration
8	Future Orientation	Conservative	Social Recognition
9	Self Accomplishment	Group Achievement	Friendship
10	Money	Success	Trust

Source: Elasmawi and Philip R. Harris. *Multi cultural Management: 2000*

USA: 'American culture' places emphasis on independence, individual freedom regardless of age, social status, or authority. Time is 'money'. They want quick responses with greatest possible profit; prefer informal approach.

Japan: The 'Japanese' value 'high quality' and give much importance to age/seniority. Besides, the Japanese wait for 'best possible results' and attach greater importance to long-term oriented approach; They have 'group devotion' and are 'conservative'.

Arab Countries: In 'Arabic cultures' also 'quality' is valued more than getting fast results. But they are particular about 'trust' in business transaction; exhibit spirit of 'hospitality and friendliness'. Success is measured by social status, reputation, honour, etc.

In view of such cultural differences, the global strategic planner should be in a position to communicate and work with people, who have been socialised in different cultural environments.

Several leading management experts have stated: "To create opportunities for collaboration, global leaders must learn not only customs, courtesies, and business protocols of their counterparts from other countries, but also have a clear understanding of their national character, management philosophies and mindsets of the people."

Fine tuning Management

The 21st century modern organisation could hope to achieve spectacular success only by furthering the interests of all its stake holders.

Stake holders are individuals as well as organisations, encompassing 'employees, customers, suppliers, distributors, creditors and associations' who would directly or indirectly exercise influence over the firm's effective operations.

As such, management's foremost goal should be to 'create value' and promote the interests of all its stake holders in a sophisticated manner. This could be achieved by commitment to certain principles.

    (i) Employees making worthwhile contributions with 'deep personal involvement' should be well-honoured for their 'excellence in spirit'. Besides, a global firm can hope to outpace its competitors in 'boosting worker productivity' by announcing greater rewards, job options and by providing best possible protection from the 'downsizing nightmare'

    (ii) The services provided by the firm should produce a 'positive emotional glow' on customers, and thereby it could enlist 'greater loyalty'.

    (iii) Announcing a generous profit sharing plan to employees in addition to salaries will enhance productivity.

    (iv) When a leader is known for his credibility and sets an example of 'high ethics', he can hope to reap substantial rewards. Here, the 'profit motive' and 'social responsibility' would coexist and enhance 'prosperity' for several sections of the society.

Concluding Observations

The 'strategy' of the modern global firm should be to acquire the ability to turnout *superb customer oriented products/services* in addition to achieving *superior productivity*. This could be made possible by employing brain-powered workforce and by making the work-place an *'enjoyable experience'*.

    The firm should avoid 'short-term gains,' 'quick fix solutions' and 'down-sizing' – which may hamper its growth on healthy lines in the long run. The 'productivity engine' of the global firm should be constantly fed by the 'creative ideas' of the technically smart people. Appropriate training modules/productivity

Strategy for Excellence in a Borderless World

improvement programmes should facilitate competent staffers to enter *the star performer ranks*.

Besides, the *team spirit* should also be constantly rejuvenated with 'useful interactions' and 'brain-storming sessions'

The global firm should make every possible endeavour to 'consistently delight' internal as well as external customers and thereby, garner the greatest accolades for its *unique products/services*.

Bibliography

1. Ansoff, H. I., *Corporate Strategy*, McGraw Hill
2. Abdul Kalam APJ, and Rajan Y.S., *India 2020: A Vision for the New Millennium – 1998*
3. All India Management Association and Amexcel Publishers Pvt. Ltd. *Strategies for Competitiveness*
4. Brown, S., *'Manufacturing the Future' Strategic Resources* (Financial Times Book: 2000)
5. Brown, S. Blackmon, Cousin, P. and Maylor, *Operations Management* (Butterworth Heine Manne, 2001)
6. Boznak, G., Rudolph and Decker, Andrey K., *Competitive Product Development:* Oxford University Press, 1993.
7. Bartlett, A and Ghoshal, S., *Managing Across Borders*, Cambridge Harvard Business Press, 1989.
8. Kennedy, Carol, *Guide to the Management Gurus*, 2002
9. Cartin, Thomas J, *Principles and Practices of Organizational Performance Excellence*
10. Deming, W.W., *Quality and Competitive Position:* MIT Press, Boston, USA.
11. Drucker, Peter, *The Practice of Management*
    - *The New World According to Drucker*
    - *Innovation and Entrepreneurship*
12. Dorling and Kindersley Publication: 2002, *Successful Managers – Handbook*
13. The "Economist" - Publication, 1999: Economics: *Making Sense of the Modern Economy*

14. Fulmer, William E., *Shaping the Adaptive Organisation*
15. Garrat, B, *The Learning Organisation* Fontana, London.
16. Grant, G.M., *Contemporary Strategy Analysis: Blackwell* Oxford, 1991.
17. Naroola, Gurmeet, *The Entrepreneurial Connection* (2001): Tata McGraw Hill.
18. Harvel, Gary and Prahalad, C.K., *Competing for the Future* - Boston: Harvard Business School Press.
19. Hammer, Prof. Michael, and Champy, Prof. James, *Re-Engineering the Corporation*, New York: Harper 1993.
20. Henk, W. Volberda, *Building the Flexible Firm*, Oxford: 1998.
21. Hill, T. and West, Brook, R., *SWOT Analysis - It is Time for a Product.*
22. Jeannet, Jean Pierre, *Managing with a Global Mindset*, (2000)
23. Adair, John, *Effective Innovation*, 1996: PAN Books
24. Johnson, G, and Scholes. K., *Exploring Corporate Strategy.* Prentice Hall, London
25. Juran, JM, *Quality Control Hand Book.* McGraw Hill, New York, USA.
26. Kanter, Dr. Rosabeth Moss, *Thriving Locally in the Global Economy 2000.*
27. Kennedy, Paul, *Preparing for the 21st Century.* New York. Random House: 1993.
28. Kotler, *Marketing Management Analysis, Planning and Control.* Prentice Hall 2002.
29. Magretta, Joan and Nanston, *What Management Is*, Free Press, 2003.
30. Marquart, Michael, and Dean, Engel, *Global Human Resource - Development.* Englewood – Cliffs NJ – Prentice Hall, 1993.

31. Maslow, A., *Motivation and Personality*. Harper and Row: New York, 1987.
32. Majumader, Ramanuj, *Product Management in India*. Prentice Hall, 1998.
33. Mintzberg, Henry, *The Rise and Fall of Strategic Planning*, New York, The Free Press, 1994.
34. Naisabitt, John, *Mega - Trends: 2000 - Ten New Directions for 1990s*, New York, William Morrow and Co., 1890.
35. Nonaka, I, and Takenchi, H, *The Knowledge Creating Company*, Oxford University Press, 1995.
36. Ohmae, Prof. Kenichi, *The Mind of the Strategist*, – *The Borderless World*. Tata McGraw Hill - 2002
37. Pande, Pete and Holpp, Larry, *What is Six Sigma*, McGraw Hill, 2002.
38. Pascale, R., and Athos, A., *The Art of Japanese*, New York 1981: Simon and Schster.
39. Paul, Trot. J. *The Innovation Management and New Product Development*. Prentice Hall, 2002.
40. Choudhury, Pran, K. *Successful Branding*, University Press.
41. Peters, T, and Waterman, R. *The Circle of Innovation*. (1997) Alfred Knoff, USA.
42. Peters, T., and Waterman, R. *In search of Excellence*, Harper Row, New York.
43. Porter, Prof. Michael E., *Competitive Strategy*, 1980 *Competitive Advantages*. New York: Free Press
44. Chadha, Rajni, *The Emerging Consumer*, New Age and Wily Eastern Ltd.
45. Senge, Prof. Peter, *The Fifth Discipline*.
46. Siddharthan, N.S., and Rajan, Y.S., *Global Business, Technology and Knowledge Sharing*, MacMillan.
47. Stacey, Ralph., *Managing CHAOS: Dynamic Business Strategies in an Unpredictable World*. (1992) Kogan Page

48. Stall, Michael J, *Management – Total Quality in a Global Environment*, Blackwell Business, 1995.
49. Toffler, Alvin, *The Third Wave*, William Collins and Sons, Great Britain: 1980.
50. Galbraith, Prof, *The Affluent Society*, Houghton Miffin, Boston.
52. Bennis, G. Warren, *On Becoming a Leader*, Perseus Press, 1994.
53. Jonathan, Cagar, and Crag. M:Vogal, *Creating Breakthrough Products*.
54. Wesley, B. Truit, *Business Planning*, Quorum Books, 2004.

Index

ABB, 170, 179
AT and T, 205
Adair, John, 121
Advanced countries (nations), 2, 6, 51, 177
Advertising, 102
Africa, 2, 191
   per capita income in, 5
Air Force Tactical Air Command, 117
Air travel, 44
   cheaper, 49
Allied Signals, 114, 138, 179
America (*see also* USA), 87, 173
   Latin America, 2
   North America, 82
   South America, 86
Ansoff, father of concept of corporate strategy, 94-95
   strategic decision-making, practical ideas to role of, 94-95
Arab countries, 222, 223
Armstrong, Gary, 82
Asia (Asian countries), 86, 167
   East Asian countries, 2
   foreign direct investment in developing countries of, 6
   per capita income in, 5
Assets, 34-35
   intangible, 168-70
   knowledge based intangibles, 32, 168-69
   physical, 32
   transferring into intangible capital, 169-70

BMW, 97
Baldridge assessment, 39
Beauty aids, 86
Beckman Instruments, 131
Beethoven, 12
Behaviours, 183
   behavioural changes, 93
Belbin, Meredith, 71-72
Belbin's team approach (team role theory), 71-72
   roles which contribute to form ideal management, 71-72
Bell Curve, 139
Bell Laboratories, 48
Bench markings, 30-31, 64-66, 137-38
   certain important attributes, 64-65
   competitive technique, 220
   factors, 220
   concept, 137-38
   essential to achieve total quality success, 129
   implementing, 65
   internal, 65
   need for, 14-15 (*see also* Competition)
   concepts, 14
   practices, successful implementation of, 15
   objectives, 16
   process, 14
   objectives, foremost, 205
   practice, 30-31
   benefits resulting from, 30
   concept of, 31
   measurement standard, 30
   competitive advantage, 30
   reduction in operating expenses, 30
   programmes, 31
   promotes effective change in organisation, 30
   use of, 220
   factors, 220
Bennis, Warren, 153, 206
Benz, Mercedez, 12
Bio-technology, 8
Blackmon, 98
Boeing jets, 80
Borderless world, 7, 171-25
   strategy for, 217-25
Bossidy, Lawrence, 114

Index

Boznak, Rudolbh G, 215
Braun, 191
Brazil, 82, 85, 86
Brown, Steve, 98
Budgetary system, 24
Business, global, 12, 13, 52, 53-54, 61, 63
  behaviour, 86-87
  budgeting, 145, 152
  co-partnerships, 15
  environment, 177
  executives, 12
  forecasting, 145
  healthy growth trends, 35
  inducting improvements, 176-77
  significant changes taking place, 176-77
  internationalising, 81
  model on core business elements, 78
  operational decisions, 209
  paradigms, 67
  planning, 145, 152
  processes, 38
  boosting employee productivity, 38
  improving efficiency of key activities, 38
  proficiency in specific core business, 168
  purpose of, 112
  pyramidal values of, 90-94
  reinvent business, 58
  shortcomings, 81
  staffers (staffing), 145, 149, 152
  strategy, 40-41, 198-99
  success and prosperity of, 69
  traditional, 53
  uncertainty in, 144-70
  assets, intangible, 168-69
  transferring into intangible capital, 169-70
  leaders as catalysts, 170
  companies, characteristics of highly successful, 158-59
  double loop or open loop learning, need for, 156-58
  Downsizing
  advantages of, 156
  risk of, 155-56
  executive excellence through organisational learning, 150-55
  Helix of never-ending improvement, 160-62
  individual attention, need for, 145-47
  knowledge base, decision making and sophisticated, 165-67
  inducing distinctive capabilities, 166-67
  leadership, special competencies required for, 162-63
  mergers and acquisitions, 167-68
  new mindset, need for, 147-50
  policy deployment, 158-60
  SWOT analysis, 163-64
  SWO findings, documenting, 164-65
  talente personnel, importance of, 155
Buyers, 116
Buzzel, Prof, 20

CAD (*see* Computer)
CAM, 172
CEO (Chief Executive Officer) (*see* Executives)
Cab, 82
Camera
  video camera, 45
Canon, 22
Capability
  to mobilise talents, 4
Capital
  capital movements, 2, 7
  continual flow of capital movements, 7
  intangible, transferring intangible assets into, 169-70
  private capital flows, 6
Carbon fibre technology, 46
Carnell, 140
Catalysts
  leaders, 170
Cellular manufacturing, concept of, 73
Champy, Prof, 204
Chassis, 82
Chile, 85
China, 153
Chrysler, 63
Clinton, Bill, 54
Clothing store
  functions of, 79 80
  augmented level, 79
  expected levels, 79
  generic product, 79
  potential level, 80
Collective brain power, 17
Collins, David J, 40, 159
  Creating Value: Successful Business Strategies by, 40
Command and control, 92

Communication, 39, 67
  facilities, 80
  home communication centres, 49
  network, 172
  revolution in, 47-48
  skin, 48
  technologies, 39, 56
Company worker, 71
Competency (ies)
  core, 197-98
  star model, 198-201
  identifying: competency mapping, 184
  approach, 184
  interviewing, 184
  questionnaire methods, 184
  twelve global, 183
Competition, global (Global Competitiveness), 3, 4, 31, 33, 65, 66, 67, 69, 80-81, 114, 217
  achieving competitive advantage and success, 107-111
  cost leadership, 107-08
  differentiation, 108
  factors, 109-11
  focus, 108
  competencies and capabilities, 19, 20
  Completer-Finisher, 72
  competitors, 83, 93, 107
  activities of, 11, 26
  best competitor, 63
  customers' competitors, 165
  direct, 65
  foreign, lower cost, 37
  latent, 65
  leap frog competitors, 65
  competitive areas, 33
  competitive activities, 11, 26, 202
  competitive advantages, 16-17, 43, 108, 109, 154
  techniques
    by decreasing cost of production, 109
    by improving quality, 109
    by rapidly responding to customers needs, 109
  competitive contributions, 71
  competitive edge, 14-15, 91
  competitive edge by healthy firm, 4
  competitive environment, 21, 165
  competitive scenario, 42
  competitive strategy, 40

competitive-success, 42
competitive 21st century, 18
consumer satisfaction and endeavouring to exceed consumer expectations ensuring, 19
core competencies, 117
cost competitiveness, achieving: implementation of strategy, 37-38
strategic steps, 38
delivery time to gain edge over competitors, reducing, 78
dimension, 217
economic, 38
factors to be considered when competing in global market, 219-20
carving out Niche Strategy, 219-20
constant review of mission and strategy, 220
constituting think tank, 220
market dominance, 219
product differentiation, 219
use of benchmarking, 220
importance, 18
managing complexity and competitiveness, 185-91 (see also Managing complexity)
methods of, 103
need for achieving competitive superiority, 14-17
benchmarking, need for, 14-15
factors, important, 16
flexibility in approach and intangible benefits, 16-17
place, producing at right, 19
product, producing right, 19
requirements, classification of, 19
sustainable competitive advantage, 19
technologically competitive market, 110
time, producing at right, 19
world class quality competition, 116
Complementary contributions, 71
Complex-adaptive system, 62-63
Computers, 7, 53, 203
Computer Aided Design (CAD), 59, 172, 203
  computer information, 49
  computer integrated systems, 56
  personal, 45
Conferencing
  teleconferencing, 178

Index

video conferencing, 178
Consultants, 65, 102
Consumer, 44, 45-46, 51, 93, 117, 179
  alternatives, 21
  consumer revolt in Japan, 45
  consumer delight, 116
  consumer focus, 69
  consumer preferences, 86
  consumer satisfaction, 116
  customer driven operational strategy, 29
  ensuring satisfaction and endeavouring to exceed consumer expectations, 19
  feed back mechanism, 28
  identification of, 52
  industrial consumer oriented services, 48
  needs, 34
  psychological factors affecting potential consumers, 27
  taboos of, 86
  tastes of, 59
Consumption patterns, rapid changes in, 116
Corruption, 81
Cosmetics, 62, 86
Costs
  cost evaluation, 38
  cost leadership strategy, 107-08
  cut down, 91
Country-uncompetitive, 6
Cousins, 98
Craftsmen, 33
Creativity (Creative excellence), 10-14, 110, 209-10
  building teams to foster, 12
  characteristics, 10-11
  creating excellence in business world, 12-13
  characteristics, 13
  creative work culture, 13
  depends mostly on adventurous entrepreneurs, 12
  business executives who induce creativity, 12
  enthusiastic innovators, 12
  leaders who ignite creativity, 12
  expertise, 10
  intrinsic motivation, 11
  knowledge, 10
  management style with accent on creativity, 13-14
  special skills, 10

Crosby, Philip, 99, 119-23 (*see also* Quality: Crosby)
  approach of, 119-23
Culture, 183
Cultural mistakes, 183
Cultural environment, 86
  functioning in global multi-cultural environment, 221-23
  multi-cultural competencies, 222
Customers, 42, 63, 98, 107, 115, 117, 179, 185, 219, 223
  building chain of customers, 72
  categories of potential customers, three, 68
  competitors' customers, 68
  constitute life blood of business venture, 68
  current customers, 68
  customer behaviour, 168
  customers' competitors, 164
  customer complaints, 77
  customer defined standards, 78-79
  customer delight, 157
    accent on, 77-79
  customer driven operational strategy, 69-73
    Belbin's team approach, 71-72
    Ricard J Schonberger's approach, 72-73
  customer driven performance, 72
  customer driven quality, 133
  customers' expectations, 68, 157
  customer focus, 202
  customer's four basic wants, 72-73
  customer pleasing limits, 139
  customer requirements, 115
  customer retention, 75-77
  customer satisfaction, 77, 95
  customer segments: classification, 68
  Customer service, 23, 117
    strategy, 157
  defined type of customer, 67
  definition of, 68-69
  delighting the customer, 69
  delivery time to gain edge over competitors, reducing, 78
  diversification in product line, taking up appropriate, 78
  essential requirements of customers, 74
  external and internal customers, 69
  former customers, 68

generate income required to carry on business, 68
greater interaction with potential customers, 28
high level of customer service and satisfaction, 17
how model is helpful, 77-80
improved frame of reference, 73-74
internal customer chain, 73
Kano's model on customer, 76-77
loyal customers, 77
matchless customer service, invigorate with, 73-75
Japanese ideology, 74-75
needs, 37, 109
new characteristics acquired, 116
new sophisticated technologies to suit latest requirements, induction of, 78
requirements, 202
specific customer type, 69, 70
status, 68
superb customer oriented products/ services, 224
superb customer service, provision of, 78
tailor made products, capability to offer, 78-79
tastes, 37
those who use substitute products or services, 68
unique services, 225
voice of customer, capturing, 78

Dankbarr, Dr, 97
Decision-making, 190-91
Delivery system, 57
Dell, 170
Deming, W Edward, 96, 99, 100, 119, 130, 159
approach of, 123-25
principles for transformation, 127
philosophy, main, 127
quality gospel, 97
Designs, 9, 51, 79
Developing world (countries; nations), 2, 6, 9, 18, 42, 44, 49, 50, 53, 115
Diffusion, need for, 43, 51 (see also Technology)
Digital Equipment Corporation, 205
Disney, 61
Downsizing, 224
disadvantage of, major, 156
risk of, 155-56
Dream teams, 12
Drucker, Peter, 7, 57,169, 170
Dupont, 29

E-Commerce, 7, 48, 58
E-mail, 48, 178
Eastman Kodak, 21-22
high-tech growth strategy, 21-22
mind-boggling plans, 22-25
Economy (Economic), global, 2, 3-4, 181
capacity constraints, 53
developing, 3-4
domestic economies, 90
economic administrators, 2
economic affairs, managing, 8
economic competition, 38
economic factors, 27
economic field, 3
economic policy, 8
economies of scale, 87
economists, 52
growth (development), 1, 2, 3-4, 50
healthy economies, 2
highly industrialised economies, 85
industrialising economies, 85
integration with global, 2, 3
greater, 3
issues, challenging, 2-3
knowledge economy, 102, 103
macro economic factors, 155
macro economic implications, 2-3
questions, 2-3
macro economic picture, 27
macro economic policy makers, 2
market economies, 6
new economic paradigm, need for, 52-53
viewpoints, 53
planned economies, 6
policy measures, appropriate, 8-9
factors to be given accent, 9
principles, 9
shortcomings to be avoided, 9
sustainable economic growth rates, 8
Efficiency, 17
Egypt, 85
Electronics, 82
electronic media, 58

Index

electronic newspaper, 49
electronic revolution, 50
Elizabeth-2, 12
Employees, 26, 39,69, 70, 77, 93, 94, 102, 109, 109-10, 156, 223, 224
   involvement in company, 134
   motivated and committed, 24
   should be trained, 161
Employment, 52
   opportunities, 1, 9, 49
Engineering
   concurrent, 203-04
   re-engineering, 203-04
   engineers, 13, 14
Entrepreneurs, 27-42, 45, 51, 59-60
   adventurous, 12
   entrepreneurial culture, 167
Enterprises, 14, 15
   function of, 64
Environment, 18, 19
   environmental degradation, 2
   environmental opportunities, 26-28
   functioning in global multi-cultural environment, 221-23
   study of, 83-87
Equilibrium, 62-63
Equipment
   new installed, 110
   suitable, 39
Ethical practices, 113
Europe (European countries), 81, 82, 86, 132
   firms, 21
Executives, 12, 13, 14, 102, 221
   Chief Executive Officer (CEO) (*see also* Executive), 4, 14, 15, 16, 19, 21, 24, 39, 58, 59, 68, 89, 90, 94, 114, 144, 145, 148, 149, 150, 151, 155, 156, 160, 163, 165, 166, 170, 174, 176, 177, 180, 183, 185, 206, 213, 221
   need for drastic changes accelerating executive talent, 214
   role of, 64
   role to tackle challenges, 178
   senior, 133
   should possess precise and useful information covering aspects, 57
   top executives, 117, 149
Expertise, 10, 32, 61
Experts, 56, 88, 92, 102, 154, 217, 223
Exports (*see also* Trade), 3, 6, 85, 88, 92
   exporter, 83

Fads, 117
Fashions, 117
   design, 79
   trends, latest, 34, 59
Favouritism, 81
Feigenbaun, 119
   benchmarks essential to achieve, 129
   contribution for total quality control, 128-29
   appraisal costs, 129
   failure costs, 129
   prevention costs, 129
   phenomenal success factors, 129
Fibre optics, 49
Finance
   financial allocation, sufficient, 27
   financial flows, 3, 80
   financial markets, 5
   financial objective, 40
   financial safety, 9
   financial soundness, 9
   research finance, 89
   strong financial systems, 2
Firms (Companies), global, 18, 19-20, 42, 52, 58, 59, 63, 64-65, 66, 68-69, 70, 74, 80-83, 115, 117, 118, 217-19
   aim of, foremost, 31
   ambitious, 16
   Cadillac, 61
   capability to make strategic choices by determining products, 21
   challenges being confronted and decisions to be taken by, 83
   characteristics of highly successful, 158-59
   competencies and capabilities, 19, 20
   competencies in specialised areas, 33
   competitive, 18
   competitive edge by healthy firm, 4
   coordinations worldwide basis, 82
   Corning, 31
   cost components, 37-38
   cover factors, 28-29
   customer service and satisfaction, high level of, 17
   decision makers, 23
   decision making process, 19
   decisions pertaining to issues, 218-19
   firm's mission and where it wants to be, 218

what action is required, 219
where firm stands now, 218
domestic, 81
dreams of success, varied, 16
effective functioning of, 21
 achieving superior performance, 21
 creating value, 21
 employees should be trained, 161
 factors be accorded priority by, 149-50
 financial strength, 28
 flexible production system, 19, 20
 flow latest information to, 21
 foreign, 81
 funds, allocation of adequate, 28
 future profitability, 37
global company : term, 82
global firm: term, 82
global operations opt for, 81
global strategic orientation, 20
global talent pool, 217
going international, 87
healthy growth of, 16
ideal system model, 70, 71
ideals and ideologies necessary for, 179
identifying potential opportunities, 34-36
 core competencies, 34-35
 market opportunities, 35-36
identifying strengths as well as resource capabilities of, 32-33
 assets, intangible, 32, 35
 assets, physical, 32, 34
 competitive edge, possessing, 33
 expertise, 32
 human factor, 32, 34
 organisational factors, 32, 35
 skills, 32
identifying threats posed, 36-37
 factors, 36-37
identifying weaknesses/deficiencies, 33-34
 internal, 33-34
implementation of strategy: achieving cost competitiveness, 37-38 (*see also* Competition)
incentives, 28
inherently possess unique knowledge, 168-69
intangible assets, 17
intellectual property of, 13
 internal strengths and weaknesses, complete analysis, 28-31
knowledge and capability tools, 17
leadership qualities, 17
learning from failures, 16
long-term benefits, 16
long-term goals, 16
mission to achieve objectives, 95
modern firms-strategic management, 25-26 (*see also* Management; Planning)
modern technologically intensive firm, 60
multifarious advantages accruing to global firm, 82
multinational and global organisation, comparison of, 188
 functional integration, 188-89
multinational company and global firm, characteristics of, 187
 objectives, 16
objectives and plans of manufacturing for world class standards, 39-41
 adherence to high quality performance in all areas, 40
 appropriate system and procedures for planning controlling and monitoring, 40
 excellent product design, 39
 inducing flexibility and reducing all rigidities, 40
 manufacturing vs buyer decisions, 40
 people and their training requirements, 40
 selecting suitable suppliers, 40
 suitable equipment and process, 39
operations on worldwide basis, 82-83
opportunity staircase for, 27
opportunities, new, 20
options, 88
organisational structure of modern, 4
periodic assessments, 39
planning, 82
 importance of, 22-25
 formal steps involved, 23
platform decisions, good, 20
principles, 95
promotions, 28
qualitative and quantitative measures, 16
reinvented, 101

Index

reinventing the firm, 64
relationship with suppliers, harmonious, 17
resources, of, 28
  core competencies, 29-30
  intangible assets, 29
  resource base, constructing, 37
  resource capabilities, 37
  tangible assets, 29
rival firms, 37-101
short-term benefits, 16
should avoid
  down sizing, 224
  quick fix solutions, 224
  short-term gains, 224
should incorporate certain principles in formulation of suitable strategies, 69
should make concerted endeavours towards, 38
six major decisions, 83
strategy, 19, 20, 21, 67, 175-76, 190, 224 (*see also* Strategy)
strategic behaviour of modern, 4
strategic control system, 38-39
steps, essential, 38-39
strategic decision matters, 28
strategic staircase, 110-11
success depends on factors, 110-11
tailoring the most appropriate strategy, 37
targets for global firm, 64, 65, 78-79
tenets to follow in order to survive, 4-5
  constant upgradation, 5
  core competencies, 4, 5
  improvements, 5
  making strategic-alliances, 5
  mergers/acquisitions to tackle challenges, 5
  outsourcing, 5
  quality standards, 4
  strategic plan, 5
  twin advantage to acquire, 61
  world class quality competition, 116
  world class status, in-depth analysis of factors, 217
Fish farming, 8
Fisher, George, 21-22
  Kodak's high tech growth strategy formulated by, 21-22
Fitz, F S, 93

Food
  retailing, 45
  super food firms, 62
Ford, 31, 61, 82, 117, 118, 119
Ford, Henry, 43-44
Foreign exchange, 8, 37, 83 (*see also* Trade)
  exchange controls, 83
  exchange rate policies, 8
  foreign direct investment, 6
France, 51, 82, 86, 173
Fuji, 22

Gardener, John W, 194
Garher, Dr, 13
Garret, Bob, 157
Geared parts, small, 82
General Electric Company (GEC), 14, 23, 87, 133, 138, 141, 142, 151, 163, 170, 179, 191, 205
General Motors, 17
Germany, 31, 51, 61, 82, 86, 118
German BMWs, 80
German Vocational Education System, 51
  massive system of technology information diffusion installed by, 51
Gestalt Therapy, 93
Gillette, 80
Globalisation
  certain major changes made by, 171-73
  challenges, 7
  global mindset, 15
  global player, requirements of, 178-80
  meaning of, 3-4
  pathways and leap frogging, 182
  process, 171-80
  gains, 1
  term, 171, 172
  think global and act global, 17
Governance
  good, 9
Grove, Andy, 112, 150

HI-IQ NET, 48
Hall Mark Card Co, 204
Hamel, Gary, 100, 197
Hammer, Michael, 204
Handy, Charles, 112
Hardware, 53
Hawthorne Effects, 118
Hay Group's study, 113

Herman, S M, 93
Hershock, 191
Hewlett Packard, 14, 22, 31, 119, 131, 132
Homes
 sophisticated insulated, 49
Honda, 14
Housing
 modern homes, 44
Human interactions, healthy, 58
Human resources, 26, 201
 activities, 28
 galvanising, 191-96
 human motivation, role of, 194-96
 categories, 195
 human resources development, 183, 191
 challenges, 183
 model of process, 193
 human resources management, 192-93
 training programmes, 192-93
 methods, 193-94
Humble, Prof
 on gaining strategic advantage, 56-57
Hungary, 6
Huxley, Aldous, 12
Hydra headed growth, 4

IBM, 17, 63, 117, 170
IMF, 9
INTEL, 170
Ichiban (number one-the biggest and the best), 74
Ideas, 102
Imagination, 102
Imai, Masaaki, 105, 106
Implementation
 accuracy, 28
 ingredients, 28
 speed of, 27, 28
 successful, 28
 timing, 28
Implications, 7-8
Imports, 3, 84, 85 (*see also* Trade)
Improvements, 103-07
 breakthrough improvements, 103, 104
 features, 107
 continuous improvements, 103, 105-07
 features, 107
Income
 high income countries, 9
 low income countries, 9, 10

national, 85
per capita income in Asian and Africa countries, 5
India, 85
Indian silk sarees, 80
Individual attention, 145-47
Indonesia, 173
Industry (ies), global, 51, 65
 analysis of, 26
 associations, 26
 changes in, 103
 changes in factory size: options for smaller plants, 49
 cost, 50
 in realms of challenges, 50-51
 industrial consumer oriented services, 48
 industrial development, rapid, 18
 industrial economies, 45
 industrial growth, 53
 industrial revolution, 7
 industrial scientists, 13
 information/experience pertaining to industrial structure, 168
 innovations, 50
 large scale, 51
 manufacturing units of future, 45
 medium scale, 51
 new, 7
 new knowledge based, 8
 productivity, 50
 quality, 50
 small industrial manufacturing units (plants), 45
 small-scale, 51
 term, 82
 timespan, 45
Inflation, 52, 81
 inflation rates, 8
Information, 102
 age of, 153
 availability of, 50
 feedback, 59
 flows, 32, 38
 revolution, 7
 systems, 57
 video text, 49
Information technology (IT), 7, 58, 174, 201
 Classic Command and Pyramid model, 56
 fibre-optic: new technology of transforming information, 49

Index

force of, 53
IT-consultancy business, 141
IT-investments, 176
IT managers, professional, 54
IT revolution in, 115
IT specialists, 176
IT systems, 56
IT-web, 57
Infotainment-vehicles, 61
Ingersoll Rand Tool, 179
Innovation (s), 16, 30, 50, 51-52, 61, 91, 103, 110, 162, 170, 209-10
  complex, 62
  drivers of, 51
  enthusiastic, 12
  hyper 61
  innovative capability: world class operations, 102-03
  changes, 103
  dimensions, 103
  intellectual capital, 102-03
  innovative culture, 167
  innovative ideas, 14, 34, 42
  innovative methods, 110
  potential for in manufacturing, 59
  process, 60
  product and process, 103
  role of, 100-01
  technical: classification, 43
  technological, 109, 174
  world class, 100-01
Intangibles
  importance of, 112-13
  qualities, 113
  stuff, 102
Integration, process of, 5-7
Integrity, 121
Intellectual
  intellectual capital, 4, 17, 102-03, 165
  directors of, 103
  horizons, 7, 57
  power, 69
  property, 102
Internet, 7, 47-48, 58, 173, 178
  Internet boom, 173
Interviewing, 184
Inventory system (inventories), 117
  cut down, 91
Investment (s), 9, 19, 38, 103
  decisions, 9

foreign direct, 6
Italy, 86

JIT, 204, 214
Japan, 31, 59, 61, 82, 84, 86, 87, 96, 97, 99, 105, 118, 123, 130, 137, 154, 159, 160, 204, 222, 223
  consumer revolt in, 45
  firms of, 21
  ideology, 74-75
Jeannet, Jean Pierre, 181
John Deere, 163
Johnson, Mike, 91
Jones, 97
Juran, Joseph, 99, 119, 130
  approach in quality control: contributions, 129-31
  ten steps to continuous quality improvements, 131
  quality trilogy of: areas, 130

Kaizen, 105, 106, 134
Kano, Noriake
Kano's model on customer, 76-77
Kay, John, 77-78, 100
Knowledge, 4, 17, 50, 56, 57, 101, 102, 169
  better knowledge-exchanges, 145
  chief knowledge officers, 103
  decision making and sophisticated knowledge base, 165-67
  essential knowledge pertaining to, 165-66
  knowledge economy, 103
  knowledge management, 103, 170
  knowledge managers, 38
  knowledge workers, 50, 58, 102
  new, creation of, 109
  organisational knowledge, 103
  worldwide awareness to treat as an asset, 172
Kodak, 63, 204
Kodak camera, 80
Kotter, John P, 82, 152, 208

Labour
  labour force, 3
  labour intensive production, 115
  labour issues, 26
  labour relations, 28, 45
  strikes, 45

Lasers
  offering vast potential, 46-47
  laser cutting, 47
  laser heat treatment, 47
  laser welding, 47
  rapid proto-typing, 47
  provide unique advantages, 47
  eliminating finishing operations, 47
  greater material utilising, 47
  higher productivity with reduced cost, 47
  superior product quality, 47
Leadership (Leaders), market, 15, 34, 39, 152, 153-54, 224
  as catalysts, 170
  exceptionally smart, 205-14
  charismatic leaders, 208
  fostering innovation and creativity, 209-10
  management and, 208-09
  need for drastic changes: ascertaining executive talent, 214
  need for tackling leadership shortage, 213
  on generic characteristics, 210
  research findings on management processes, 212-13
  split brain theory and effective management, 211-12
  left hemisphere, 211-12
  right hemisphere, 211
  inspiring, 22, 170
  management leadership, 121, 208-09
  principal tasks of, 152
  qualities, 17, 153-54
  special competencies required for, 162-63
  visionary leadership, 51
  with greater involvement, 133
Lean production, concept of, 54
  term, 54
Leap frogging globalisation pathways, 182
Learning, 30
  double loop or open loop, need for, 156-58
  factors, 156-57
  strategic, 170
  systems, 183
Levitt, Tedd, 20, 79
Life

living standards, 1, 2, 3, 10
  quality of, improving, 42, 49
Logothetis, 120, 121

Macro
  macro economic implications, 2-3 (see also Economy)
  macro economic management, international, 4
  macro economic stability, 6, 9
Magnetic Levitation (MAGLEV) passenger train, 46
Maister, David, 207
Malcolm Baldbridge National Quality Award (MBNQA), 15
  core values, 15
Management, 7, 19, 24, 45, 109-10, 158
  Belbin's roles which contribute to form ideal management, 71-72
  commitment, 122
  company wide motivation, 119
  consultants, 102
  excellent project management, 161
  experts, 88, 92, 102, 154, 217, 221, 223
  fine tuning, 223-24
  integrated plan, 91
  knowledge management, 103, 170
  leadership and, 208-09
  levels of, 28
  management behaviour, 130
  management by policy, 159
  management gurus, 1
  management leadership, 121
  management processes, 92, 120
  management-pyramid, 102
  managing quality, 129
  managerial practices, 10
  managing complexity and competitiveness, 185-91
  managing uncertainty, 185-86
  objectives, multiple, 186
  operations, geographic scope of, 186-91
  decision-making, 190-91
  flexibility in approach, 189
  global norms to be followed, 189
  orthodox management models, 92
  philosophy, 115
  process management tool box, 111-12
  process, six vital factors achieving excellence in, 25

Index

project management, 119
research findings on management processes, 212-13
roles, 152-53
split brain theory and effective management, 211-12
stable normal management, 146
statistical process control, 119
strategic management process, 24-26 (see also Planning)
mission, vision and goals, 25-26
style with accent on creativity, 13-14 (see also Creativity)
systems management, 119
time management, 74
top management, 91, 160
commitment, 118
Total Quality Management (TQM), 29, 115, 216 (see also Quality)
programmes, 39
Manager (s), global, 4, 10, 11, 16, 19, 24, 39, 57, 65, 89, 92, 93, 102, 125, 141, 146, 153, 174, 177, 180
authoritarianism, culture of, 93
enlightened, 14
general manager, 14, 56, 152, 165
ignite intrinsic motivation, 11
learning from experience gained in past, 16
managerial knowledge, 129
middle managers, 56, 148
multi-cultural, essential characteristics, 222
personnel manager, term, 191
pressures, exposed to, 93
professional IT managers, 54
should have clear ideas, 160-61
should induct certain distinctive capabilities, 166-67
sophisticated judgement, 11
training managers, 122
Manufacturing, 89
Agile Manufacturing Model, 215
computer integrated manufacturing systems, 56
continuum, ideal, 54-56
manufacturers, 7
potential for innovation in, 59-62
processes, 109
technologies, 52
world class, 115, 214-17
six sigma production, 216-17
technology of process, 215-16
technology of product, 215
Market (Marketing), global, 3, 15, 63
analysis of, 26
changes in, 103
competitive market area, 50
constituting global marketing organisation, 88-89
domestic, 20
dominance, 219
experts, 88
financial, 5
forces, 51
foreign, 20
global terms, 89
globalisation, 103
integrated, 6
international, 7, 83
lead markets, 181
detection and identification of, 181
local, 7
methodology of entrance to, 88
options, 88
mix, 88
new, 16, 19, 80
opportunities, 35-36, 37, 43, 52, 87
place, winning battle in, 50-51
policies, 87
position, 40
regional, 7
segments, 29
selecting suitable markets, 87
share, 16
strategist, 35
technical market intelligence, 60-61
techniques, 91
training programme, suitable, 89
trends, 28, 202
working out global marketing programme, 88
options, 88
Maskell, Prof, 214
Maslow, Prof, 195-96
Massachusetts Institute of Technology, 54
Maylor, 98
Media car
emergence of, 61
Mega network, 48

Mental geography, 7
Mercedes, 97
Mercedez Benz, 80
Merck, 29
Mergers and acquisitions
　distinctive capabilities on enterprise, 168
　term, 167
Micro
　micro level, 4-5
　microsoft, 17
Milliken and Co, 117
Mills, 140
Mindset, global, 15, 92, 177, 183, 217
　developing, 180-84
　intoxicated-mindset, 4
　need for new, 147-50
　basic capacities, 147-48
　decisions, 148
　strategic-outfit: factors, 148-49
　personal characteristics, 181
　traditional and: comparison of, 180
Monetary, policies, 8, 53
Monitoring
　control and, 27
　monitor-evaluator, 72
Moren, Dr, 183
Morita, Akio, 51
Motivation, 152, 183
　company wide motivation, 119
Motor drives, special, 82
Motorola, 21, 34, 61, 131, 138, 179, 193, 205
Mozart, 12
Multi-disciplinary teams, 12
Multi-leveled thinking frame, 13
Multinational Corporations (MNCs), 80

Natural resources
　cobalt, 85
　coffee, 85
　copper, 85
　oil, 85
　tin, 85
　negotiation, 183
　never ending-improvement-Helix, 160-62
　newspaper
　electronic, 49
　niche strategy, 219-20
　Nokia, 61, 80
　novelty, 62-63

Omark, 205
Operating expenses, 16
Opportunities, global, 1, 9, 20, 26-28, 31, 35-36, 37, 43, 52, 87, 92, 165, 166, 217
　identifying new, 27
Optimisation, 63
Organisation, global
　building organisational capability, 92, 98-100
　continuous improvement, 99-100
　quality culture, 100
　top management commitment, 99
　total quality management, 100
　world class operations-role of quality, 98-99
　business climate, 92, 93-94
　business strategy, 92, 94-97
　critical factors, 95
　just in time production, 96
　lean-production, 96-97
　mission, 95
　principles, 96
　quality, 96
　technique, 96-97
　value stream concept, 96-97
　vision, 95
　critical success factors, 95
　decision types, organisation should concentrate on, 95
　mission to develop and implement people strategies and plans to enable company to
　achieve objectives, 95
　develop adaptive capabilities, 153
　organisational culture, 103
　organisational environment, 92, 93-94
　Gestalt approach to, 93-94
　need for congenial work environment, 94
　organisational excellence, achieving, 196-205
　benchmarking, 205
　building new capabilities: global approach, 196-98
　core competencies, 197-98
　customer focus, 202
　engineering, concurrent, 203-04
　re-engineering, 203-04
　star model, 198-201
　activities, 200-01

Index

outsourcing, 201
star model on core competencies for change in performance, 198
organisational improvements, 16
organisational knowledge, 103
organisational learning, 150-55
organisational objectives, 162
organisational operations, 15
organisational performance, 92
elements, three core, 92
organisational renewal in industry groups, 3-4
policy matters, 95
power hierarchy, 94
procedures, standard operating, 95
programmes, 95
reorganisation, 92
responsiveness, 217
strategy, 95
strategic staircase, 110-11
success depends on factors, 110-11
vision through people: extraordinary customer's satisfaction, 95
Otis Elevator, 82

Pacific Rim countries, 81
Packard, Hewlett, 208, 131, 205
Panasonic, 80
Parmerter, Dr, 13
Partisan attitude, 81
Partnership
strategic partnerships, 51
People, 169
need for individual attention, 145-47
Performance standards, 14
Perpetual novelty, 62-63
Peters, Tom, 73, 79, 95, 98, 100, 117, 155
Pharmaceuticals, 62
Philippines, 85
Photocopying, 39
Piece work basis, 49
Planning, 22-25, 66, 89
budgetary system, 24
controlling performance, 24
evolution of major planning efforts, 23
formal steps involved, 23
ideas, new, 23
implementation of plan: factors, , 24
incentive programme, 24
management, 24

monitoring performance, 24
reward system, suitable, 24
strategic planning management process, 24-26
classification, 24
control system, 25
environmental opportunities and threats, analysis of changing, 25
implementation, 25
internal strengths and weaknesses analysis of changing, 25
long-term externally oriented issues, 24
mission, vision and goals, formulation of organisations, 25-26
SWOT analysis, 25
short-term, tactical and operational issues, 24
strategic techniques, 161, 192-93
Plant (Staff), 71
Poland, 6
Policies, 2
policy deployment, system of, 159-60
Catch Ball, discussion and debate over objectives termed, 160
Koshin Kanri, Japanese term, 160
policy deployment processes, 159-60
policy framework, 3
policy makers, government mindsets of, 3
Political field, 3
Porras, 159
Porter, Michael, 58-64, 107, 108, 199-200
Poverty (Poor), 2, 9, 10
extermination of, 2
levels, 2
Power hierarchy, 94
Prahalad, C K, 197
Premji, Azim, 216
Private enterprises, 8
Problems confronted
characteristics, general, 5-7
correct problem-solving methods, 161
Process
statistical process control, 125-27
control chart, 126
Product (Production; Productivity), 50, 63, 83, 91, 109
benefits emanating from product standardisation, 20-21

developing innovative productivity, 114
differentiation, 219
focus on outer rings of total product, 79
for specific services, 73-74
generic, 79
globally standardised, 20
high standards of product excellence, 52, 113
innovative, 37, 89
labour intensive production, 115
lean production, 96-97
concept of, 54
term, 54
local customised, 20
product adaptation, 21
product and process innovations, 21
product centric thinking, 70
product designs, 51
product development, 202
product excellence, 14
product technologies, 52
productivity improvement programmes, 224-25
quality control, 91
repositioning of, 80
short product cycles, 117
shorter life cycle of several products, 60
standardisation
benefits emanating from, 20-21
design-modification, reduced costs of, 21
economies experience, faster accumulation of, 21
learning experience, faster accumulation of, 21
strategy, 35-36
cash low segment, 36
star quality segment, 36
superb customer oriented, 224
superior productivity, 37, 224
technology of, 215
two Outer Levitt Rings, 80
unique products/services, 225
unique value in, 69-70
Prosperity, 2, 14
Protectionism, 81

Quality, 98-100
action, taking correction, 122
as major strategic factor, 114
awareness, raising of quality, 122
continuous improvements: two factors, 132
cost of, evolution of, 122
Crosby, Philip, approach of, 119-23
philosophy of, 123
preventive medicine, 121
quality vaccine, 121
quantitative approach, 120, 121
salient absolutes of, 119-21
step programmes, fourteen, 121, 122-23
stages of development, five, 123
strands of thought, three, 121
customer driven quality, 133
definition of, 119, 120, 123
Deming, W Edward
approach, 123-24
deadly sins, seven, 124, 125
definition, 123
methodology, 124
problem solving, 124
theme, 123-24
development of quality culture, 127
philosophy, main, 127
principle for development of quality culture, 127
focused quality management, 134-37
achieving measurable results, 135-37
emphasis on focus, 135
factors to be considered, 134-35
four step approach, 134
Feigenbaum, contributing of, 128-29
phenomenal success factors, 129
goal setting, 122
improvement programmes, 137
inspection, 120
mass inspection, 120
Juran, Joseph
contribution of quality control, 129-31
ten steps to continuous quality improvement, 131
management commitment, 122
management: viewpoints of leading architects, 119
approaches, 119
gurus and their teachings, 119
manufacturing quality products, 117
new demands for quality consumer products, 117
new standards and strengths required in 21" century milieu, 116-19

Index

new standards for quality imposed, 110
of conformance, 123
of design, 123
of process, 116
of product, 116
of service, 116
people empowerment: four aspects, 132
preventive medicine for poor quality, 121
principles determining quality/performance-standards, 116
quality certification, 98
quality conscious work force, 18
quality control, 130
process, 129
quality costs, 128
quality councils, 122
quality culture, 100
quality improvements, 130, 131, 134
effort, 129
teams, 122
quality is free, 123
quality leadership team (QLT), 135-36
quality management, 129-31
quality measurements, 122
quality movement, 123
quality products, 18
quality revolution, 99
quality vaccine, 121
categories, five major, 121
communications, 121
integrity, 121
operations, 121
policies, 121
systems, 121
recognition, 122
reporting: errors/obstacles, 122
six-sigma quality: Bell Curve, 139
statistical process control, 125-27
control chart, 126
targets, 118
tools and techniques, 131-34
continuous improvement, 134
customer driven quality, 133
employee involvement, 133
just in time exercise, 133
leadership with greater involvement, 134
product design/manufacturing processes, 131-32
total quality control (TQC), 128-29, 132, 133
definition, 133
total quality cost, 128-29
total quality efforts, 114
total quality management (TQM), 99, 100, 115, 134, 216
philosophy, 133
training programmes for, 142-43
objectives, 142-43
results should be, 143
to achieve certain ends, 142-43
transformation, reasons for, 115-16
trilogy of quality planning, 130
truly successful quality initiative, 124
world class quality, 98-99
factors, 98-99
Quantification, 121
Quatar
Doha, WTO meeting (2001) at, 85
Questionnaire methods, 184
Quota system, 83 (*see also* Trade)

R and D (Research and Development), 13, 14, 33, 51, 58, 59, 60, 61, 82, 109, 170, 189
activities, 29
Rail roads, 7
Rajan, Amini, 158
firms: key ingredients, five, 158-59
Rank Xerox Corporation, 3, 63, 137, 163, 205
Raw materials, 85
Reagon, Ronald, 153
Refinements, effecting, 27
Regulations, 37, 81, 93, 169
regulatory issues: political/social milieu, 26
Relationships, 91, 169
firm's harmonious relationships with suppliers, 17
formula, 67
inter personal relations, 93
quality relationships, 145
activities, 29
Resource investigator, 72 (*see also* Resources)
Reven, Reg, 157-58
Reward system, suitable, 24
Rhinesmith, Stephen H, 180, 187, 188
Managers Guide to Globalisation by, 180
Riesenberger, Dr, 183
Risks, 2, 7

factors, 2, 6, 56
high-risk factors, 148
types of, 1
unavoidable, 81
Roads
  rail roads, 7
  road map, 18, 67
  safer road surfaces, 49
Roboties, 59
Rock, Granite, 91
Rolls-Royce, 12
Rover Group, 95
  success factors, 95
Royal Dutch/Shell, 163
Russel, Prof, 204
Russia, 173

SWOT analysis, 25, 31-32, 37, 163-64
  dimensions, 164
  documenting findings, 164-65
  primary internal and external conditions, 164
  stands for
  opportunities, 163
  strengths, 163
  threats, 163
  weaknesses, 163
Saudi Arabia, 85
Schonberger, Richard J
  approach of, 72-73
  customer focused principles, 73
  world class excellence, 72-73
Schumpeter, Joseph A, 101
Science and technology, 50
Scientists, 13, 14, 48, 53
Sears-Roebuck, 17, 75
Semi-conductors, 53
Senge, Peter, 153
Sewell, Carl, 75
Shakespeare, William, 12
Shaper, 71
Shaw, Bernard, 147
Siemens, 14, 31, 163, 170
Silicon Graphics, 163
Six sigma, 138-39, 140-42, 178-79, 216-17
  basic equation on, 141-42
  programme, 140-41
  standards, 34
  success of, 216-17
Skills, global, 26, 32, 34, 97, 154, 202

developing, 180-84
skilled workers, 6
special, 10
specialised, 6
technical, 13
Sloan
  behavioural study findings of, 44
Smith, Adam, 44
Social
  progress, 1, 8
  socio-economic fabric, 70
  socio-economic health, 18
  socio-political framework, 44
Software, 53, 102
  engineering, 102
Sony Corporation, 14, 51, 61, 63, 80, 117, 132, 170
Soviet Union, 6
Spain, 82
Specialisation, 3
  specialists, 89
Spendolini, Michael, 30
Sperry, Roger, 211
Sports equipment, 46
Stacey, Ralph,
  Managing Chaos by, 146
Staff (staffers), 93
  categories, 183
  competent, 225
  reduction of, 91
  star performer ranks, 225
Standard of living (*see* Life)
Statistical process control, 119, 125-27
  control chart, 126
Sterling Chemicals, 163
Stewart, Thomas A, 102
Strategy of modern global firm (Strategies: generic), 67, 175-76, 190, 224
  constant review of mission and, 220
  Prof Humble on gaining, 56-57
  strategic alliances, 166
  strategic decision-making, practical ideas formulated by Ansoff in regard to role of, 94-95
  strategic goals, 11
Super food firms, 62
Supervisions
  training supervisions, 122
Suppliers, 165, 223
Sustainable value creation, 20

Index

Sweden, 59
system (s), 67, 68

Talents (talented personnel, 4, 91, 113
    importance of, 155
Tanzania, 86
Tariff (s) (*see also* Trade), 81, 83
Tastes, 34, 117
Taxation
    tariff, 81, 83
    tax rates, 8
    tax structure, 83
Taylor, Prof, 204
Team spirit, 225
Technology (ies), 14-15, 42-46, 63, 93, 153, 176-77, 202
    advanced, 137-40
    benchmarking, 137-38
    manufacturing, 59
    need for capturing spirit of six sigma, 138-40
    changes in technologies, 103
    cheaper, 36
    communication technology, 56
    core technology areas, 102
    diffusion of, need for, 43, 51
    massive system of technology information diffusion installed by Germany, 51
    diverse, 64
    effective training in technologies, 118
    emerging new technologies, 53
    laser technology, 46-47 (*see also* Lasers)
    latest developments, 59-62
    manufacturing technologies, 50-59
    mass production techniques, 43
    mobility of, 53
    new, 7, 101
    new by 2020, 46
    sophisticated technologies, 78
    of process, 215-16
    of product, 215
    prediction of certain important changes in technological advances, 45-46
    radical changes in, 174
    rapid improvements, 172-73
    specific technologies, 50, 51
    sophisticated, 52, 78
    superior, 57, 113
    technical experts, 56
    technical ideas, 109

technical innovation: classification, 43
technical know-how, 60-61
technical knowledge, 3, 30
technical market intelligence, 60-61
technical skills, 13 (*see also* Skills)
technological abilities, 53-54
technological breakthroughs, 12
technological changes, 49
technological developments, 42, 168
technological factors, 27
technological innovations, 109
    categories, two major, 13
    codified, 13
    tacit, 13
technological strength, 42
technological upgradation, 51
technologists, 48, 52, 92
    to be managed effectively, 3
    use of, 173-75
    Michael Porter's approach, 174-75
    Michael Porter's competitive forces model, 175
Telecommunication (s), 49, 53
Teleconferencing, 39, 178
Telephone, 49, 178
    conversations, 49
    mobile-personal, greater use of, 49
    service, 44
Television
    broadcasts, 49
    multiplicity of special interest TV channels, 49
    two-way television, 49
Texas Instruments, 117, 179, 205
Think tank
    constituting, 220
Threats
    future, 166
    strategies for overcoming possible threats, 175
3-M
Toyota Motor Company, 14, 17, 61, 73, 80, 96, 97, 119, 132, 134, 170, 205, 215
    Lexus Car, 97
    Toyota cars, 80
Trade, 3, 5
    barriers, 81, 84
    embargo, 84
    exchange controls, 83
    expansion of, 10
    exports, 85, 88

foreign exchange, 83
free trade, 53
General Agreement on Tariffs and Trade (GATT), 84-85
  import, 84, 85
  international, 80-81
  trading system, 83-87
  non-tariff trade barriers, 84
  quota system, 83
  tariff, 83
  Uruguay Round of Discussions, 84
  World Trade Organisation (WTO), 9, 84-85
  agreements, 85
Training, 130, 161
  improved methods, 110
  managers, 122
  modules, 224
  objectives, 142-43
  programmes, 28
  programmes for quality, 142-43
  results should be, 143
  supervisors, 122
  to achieve certain ends, 143
Transactions, transparency in, 9
Transformation
  Deming's principles for, 127
Transmission
  revolutionary changes in, 49
Transport, 172
  transportation network, 80

UK (United Kingdom), 51, 78, 95
USA (United States of America) (*see also* America), 35, 61, 75, 80, 82, 86, 123, 124, 132, 151, 155, 181, 182, 222, 223
  American Management Association, 155
  Boston Consulting Group, 35
  management practices, 124
Unilver Company, 52
Uruguay Round of Discussions, 84

Value (s), 92
  value analysis techniques, 38
  value building offerings, 18, 40
  Value Chain Approach, 199
  Value Chain Model, 199
  value creation, 17

value culture, 26
value stream concept, 96-97
Vehicles
  quiet/less polluting, 49
Vickens, Peter
  policy deployment processes provided by, 159-60
Video
  Video taping, 39
  Video text information, 49
Volatile capital flows, 6

Waldrop, Mitchell, 62
  Complexity by, 62
Walkman, 51
Wall Mart, 17
Walmart inventory system, 75
Welch, Jack, 4, 23, 141, 142, 151-54, 163
Welch, John, 4
Western countries, 118-19, 173
  Hawthborne Effects, 118
  management practices, 124
  Western manufacturing firms, 118
  zero defect programmes, 118
Westing House, 205
Wheat, Barbara, 140
Whirlpool Corporation, 213
Whitman, David, 213
Wipro, 216
Womack, 97
Work
  work culture, creative, 13 (*see also* Creativity)
  workplace: efficiency in, 109
Worker (s) (*see also* Employees)
  company worker, 71
  migration of skilled, 6
World class excellence, 72
World Trade Organisation (WTO), 84-85 (*see also* Trade)
World truck sport, 82
World War II, 44, 181

Zaire, 85
Zero defect programmes (Zero defects), 118, 120, 121, 122, 123, 178
Zero Defect Day, 122